Semantic Interaction for Visual Analytics

Inferring Analytical Reasoning for Model Steering

Synthesis Lectures on Visualization

Editors
Niklas Elmqvist, *University of Maryland*
David S. Ebert, *Purdue University*

Synthesis Lectures on Visualization publishes 50- to 100-page publications on topics pertaining to scientific visualization, information visualization, and visual analytics. Potential topics include, but are not limited to: scientific, information, and medical visualization; visual analytics, applications of visualization and analysis; mathematical foundations of visualization and analytics; interaction, cognition, and perception related to visualization and analytics; data integration, analysis, and visualization; new applications of visualization and analysis; knowledge discovery management and representation; systems, and evaluation; distributed and collaborative visualization and analysis.

Semantic Interaction for Visual Analytics: Inferring Analytical Reasoning for Model Steering

Alex Endert

ISBN: 978-3-031-01475-8   paperback
ISBN: 978-3-031-02603-4   ebook

DOI 10.1007/978-3-031-02603-4

A Publication in the Springer series
SYNTHESIS LECTURES ON VISUALIZATION

Lecture #7
Series Editors: Niklas Elmqvist, *University of Maryland*
                David S. Ebert, *Purdue University*
Series ISSN
Print 2159-516X   Electronic 2159-5178

Semantic Interaction for Visual Analytics

Inferring Analytical Reasoning for Model Steering

Alex Endert
Georgia Tech

SYNTHESIS LECTURES ON VISUALIZATION #7

ABSTRACT

User interaction in visual analytic systems is critical to enabling visual data exploration. Interaction transforms people from mere viewers to active participants in the process of analyzing and understanding data. This discourse between people and data enables people to understand aspects of their data, such as structure, patterns, trends, outliers, and other properties that ultimately result in insight. Through interacting with visualizations, users engage in sensemaking, a process of developing and understanding relationships within datasets through foraging and synthesis.

This book discusses a user interaction methodology for visual analytic applications that more closely couples the visual reasoning processes of people with the computation. The methodology, called ***semantic interaction***, affords user interaction on visual data representations that are native to the domain of the data. These interactions are the basis for refining and updating mathematical models that approximate the tasks, intents, and domain expertise of the user. In turn, this process allows model steering without requiring expertise in the models themselves—instead leveraging the domain expertise of the user. Semantic interaction performs incremental model learning to enable synergy between the user's insights and the mathematical model.

The contributions of this work are organized by providing a description of the principles of semantic interaction, providing design guidelines for the integration of semantic interaction into visual analytics, examples of existing technologies that leverage semantic interaction, and a discussion of how to evaluate these techniques. Semantic interaction has the potential to increase the effectiveness of visual analytic technologies, and opens possibilities for a fundamentally new design space for user interaction in visual analytic systems.

KEYWORDS

user interaction, visual analytics, model steering, visualization

Contents

Acknowledgments

I appreciate the feedback and comments on drafts of this manuscript from Ross Maciejewski, Remco Chang, and Niklas Elmqvist. I would like to thank Doug Bowman, Scotland Leman, Richard May, and Francis Quek for serving on my Ph.D. committee and providing valuable feedback that helped create much of the work presented in this manuscript. Finally, I'd like to thank Chris North for his continued guidance, advice, and encouragement that made this research possible.

Alex Endert
September 2016

CHAPTER 1

Introduction

Visual analytics helps people make sense of data by combining information visualization with data analytics through interactive user interfaces. The current data-driven era emphasizes the importance of usable, yet powerful tools that help people understand their data, and thus the world around us. Visual analytic applications serve an important role in these situations, as they create usable interfaces for people to discover insights into their data.

Performing visual data analysis involves a blend of data analytics and human reasoning. The appropriate blend of these two parts needs to be explicitly balanced through design and evaluation. To achieve this balance, people and computers co-reason about data using each of their processing abilities, with the goal of giving people insights. There has been much emphasis on the computational and analytic components of visual analytic systems. The same is true of the visual side; novel visualizations continue to be created. As a result, researchers and developers have created advanced analytics that are faster and more accurate, and able to uncover more complex forms of phenomena and patterns from data. Novel visualization techniques have also been researched and developed that allow people to perceive and understand new and intricate data characteristics, as well as results of analytic models. User interaction, on the other hand, has seen fewer innovation and design alternatives.

The integration of analytic models into visual analytics continues to be an open research challenge. While the visual analytics community advocates for approaches to data analysis that include human perception and cognition, there are advantages to incorporating analytic models into the process. Some data analysis tasks are more well-suited for automation, while other tasks that are perhaps less structured and defined are better for humans to perform. For example, computing clusters from data is more efficiently performed by computation, as long as the parameters of the clustering algorithm are known by the user. In contrast, generating questions, hypotheses, or stories about potentially interesting insights may be better suited for human reasoning. The understanding of which tasks during analysis are better suited for computation or cognition is an open area of research, yet an area well-suited for visual analytics.

User interaction is a growing focus area for visual analytics. The understanding of how to integrate people into the computational processes and visualization methods used for visual analytics is increasing in importance. Interaction transforms people from mere viewers of visualizations into active participants in the visualization and analysis processes. People can contribute to this process in different and important ways. Their expertise and knowledge about a domain can allow them to focus the computation toward more relevant or important features of the data.

Alternatively, in order to understand the data being shown, exploring different aspects and views of that data might be helpful.

Semantic interaction is an approach to user interaction that couples exploratory interactions with updating and steering computational models. The premise of semantic interaction is to create user interaction techniques that more closely couple this cognitive processing and reasoning of humans with the computational processes and models used in visual analytics. The goal is to create systems that optimize the balance between human and machine effort for data analysis. For such systems, user interaction is the means through which this coupling takes place.

1.1   THE ROLE OF VISUAL ANALYTICS IN A DATA-DRIVEN ERA

Visual analytics is a science based on supporting sensemaking of large, complex datasets through interactive visual data exploration [69]. The success of such systems hinges on their ability to combine capabilities of statistical models, visualization, and human intuition—with the goal of supporting the user's analytic process. Through interacting with the system, users are able to explore possible connections, investigate hypotheses, and ultimately gain insight. This complex and personal process is often referred to as sensemaking [57].

In the digital era, an increasing number of physical phenomena are being digitized and quantified through sensors and data transformations. This data is stored in ever-growing databases, against which scientists, practitioners, and others pose questions about the world and the phenomena in it. This can reveal new insights into existing areas of science, and also help people working in other domains such as security, finance, business, and others. For many domains and tasks, the mere presence of data presents opportunities to answer existing scientific questions using new, data-driven methods.

Visual analytics occupies an important role in the myriad of data science tools that help people make sense of data. For example, visual analytics can help people understand and contextualize the results of automated computation. Often, simply finding the answer using one or more automated methods leaves people questioning the accuracy of that answer, or how much they should be able to trust it [99]. In these situations, visual analytics can help people explore the processes and decisions of how these automated processes came up with the results. Visualization is a powerful mechanism for explaining the results of artificial intelligence, machine learning, predictive modeling, and other automation. It can help people understand why models came up with the answer they did, why other alternatives were not returned, what the confidence of these results are, etc.

Alternatively, visual analytics can also serve as the user interface to help people explore and analyze data more freely. People can ask questions, test hypotheses or prior expectations about the data, and engage in other, more exploratory tasks. For the latter, the visual analytic techniques provide people with the interfaces needed to tune and adjust the analytic models that compute on the data, and are often used to generate the visualizations contained in them. Finally, visual analytics can help overcome some of the potential drawbacks in automated computation (e.g., helping machine learning models delineate signal from noise), as well as human-in-the-loop techniques (e.g., helping people formulate the questions to ask of their data) [98]. The growth and evolution of visual analytics plays a pivotal role in the advancement of data science.

1.2   SEMANTIC INTERACTION

Semantic interaction [25] is an approach to user interaction that enables coupling between computational models and human reasoning. Semantic interactions are designed to be performed on visual objects that are well-understood by analysts in the domain of the data. Sequences of semantic interactions are systematically interpreted to produce updated and steered mathematical models, which in turn drive updated visualizations and foster analytic discourse. Semantic interaction interprets sequences of user interactions as data structures that capture semantically meaningful information about the user so that systems can use the information during analysis, and also analyzing a user's process after analysis.

Visual analytic techniques that make use of semantic interaction let people interact with the visual representations themselves to impart their domain expertise onto the computational models. For example, people may have domain expertise that suggests a strong similarity between two or more data items. This can be conveyed by visually grouping those items. In response, the system learns the similarity function that can describe this relationship, and applies the learned model to the rest of the data to produce a new visualization. This iterative process continues as people explore the data and gain new insights.

To date, semantic interaction has focused on enabling direct manipulation of spatializations, which are two-dimensional views of high-dimensional data such that similarity between information is represented by relative distances between data points (e.g., a cluster represents a collection of similar information) [66]. Other applications of semantic interaction include additional visual metaphors, different computational models, and other domains. Some of these are discussed in the later sections.

The interactions people perform are small, piecewise realizations of the cognitive analysis process of analysts. Thus, *semantic interaction* is a systematic process by which the analytical reasoning or meaning (i.e., the *semantics*) of the interactions are interpreted by the application. As a result, the cognitive processes of reasoning about data are more closely connected with the computational processes of machines.

To illustrate the concept of semantic interaction, an existing prototype ForceSPIRE (shown in Figure 1.1) is used throughout the following chapters. ForceSPIRE is a visual analytic prototype incorporating semantic interaction for analysis of text document collections represented in a spatialization [25]. Semantic interactions in ForceSPIRE include repositioning documents, highlighting text, searching, and annotating documents. When users perform semantic interactions in the course of their reasoning process, the system incrementally updates a keyword weighting scheme in accordance with the user's analytical reasoning (Table 4.1). The learned weighting scheme emphasizes relevant keyword entities within the dataset and adjusts the layout of the spatialization accordingly. Thus, the goal of ForceSPIRE is to automatically steer the spatialization based on the user's interaction with a subset of the information. It is used to test and evaluate the balance between people and machines in visualizing text documents spatially, via multiple analytic models.

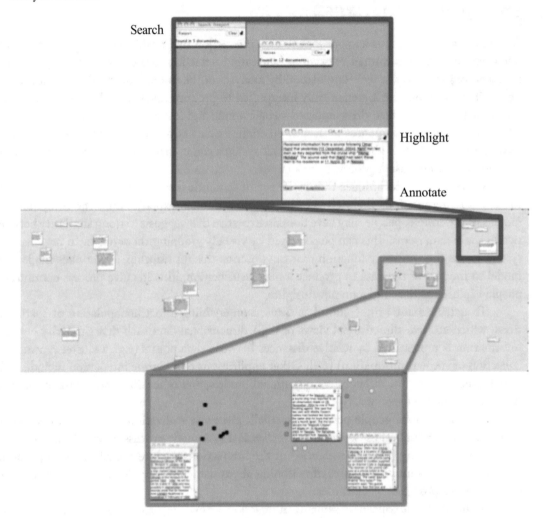

Figure 1.1: A scaled-down screenshot of ForceSPIRE taken on the large, high-resolution display used in this study (two zoomed views shown). Users can search, highlight, annotate, and reposition documents spatially. Documents can be shown as minimized rectangles, as well as full detail windows.

1.3 OUTLINE

Chapter 2 starts with an overview of the fundamental concepts that ground semantic interaction for visual analytics. It consists of topics including model steering, mixed-initiative systems, analytical reasoning, sensemaking, and creating models from captured user interactions from visualizations.

Chapter 3 discusses the value of the spatial visual metaphor, specifically for sensemaking. It covers some of the basic studies and research performed which grounds much of the spatial adjustment and manipulation of data underpinning semantic interaction.

Chapter 4 covers the design guidelines for semantic interaction. These are described in context of applying this form of user interaction to visual analytic systems with a variety of dataset and analytic models. Many of these design guidelines are exemplified in applications shown in subsequent chapters. This section also describes a general framework for semantic interaction.

Chapter 5 provides a select set of visual analytic applications that utilize different forms of semantic interaction. The coupling between the user interactions and the model steering techniques are discussed for each.

Chapter 6 presents current approaches for evaluating systems that utilize semantic interaction. The implicit model steering that is at the core of semantic interaction requires novel methods for evaluation. In addition, the captured temporal models approximating user interest opens opportunities for novel user evaluation techniques for visual analytics.

Chapter 7 discusses semantic interaction more broadly in the context of visual analytics. Additionally, it mentions some of the research directions and challenges raised. Chapter 8 provides an overall summary and conclusion.

CHAPTER 2

Fundamentals

Semantic interaction is grounded in well-established human-computer interaction and information visualization principles. These foundations are explained in more detail in this section, and used to support the information in the subsequent chapters. First, a description of sensemaking and analytical reason frame the cognitive processes and tasks that people perform with visual analytic techniques. Next, the importance of analytic models for visual analytics is discussed. A short discussion of user interaction with visual analytics is also described, including a collection of prior work illuminating that analytical reasoning can be computed from user interaction logs. Finally, the concepts of model steering and mixed-initiative systems are introduced. The content in this section is not intended to be complete surveys or summaries of the existing work, yet serve to ground the discussion of semantic interaction in the subsequent chapters.

2.1 SENSEMAKING AND ANALYTICAL REASONING

When people make sense of data visually, we make use of a wide variety of computational and visual tools and applications. Many design considerations and guidelines go into making these tools as effective as possible, ranging from visualization and interaction guidelines and principles.

However, it is equally important to recall that one of the primary reasons for producing interactive visualizations is to help people think and reason about the visualization. The design of interactive visualizations should be tailored toward the production of insight, not only the optimization of computational and visual components of the system. Below, we describe existing models that describe the cognitive processes of people performing data analysis.

One popular description of the analysis process of people is Pirolli and Card's "sensemaking loop" [57, 81], depicting stages of the cognitive process humans engage in to translate raw data into presentable insights. During this process, humans develop mental models of the information with respect to the domain of the data. They see patterns, trends, outliers, and other data characteristics and hypothesize about their relevance and importance with respect to the domain. The process entails a sequence of internalizing information, and externalizing incremental questions and knowledge constructs back into the system and environment.

An alternative description of sensemaking has been proposed by Klein et al. [82, 83]. They describe the sensemaking process in terms of "frames," where humans continually build and modify their understanding (i.e., frames) of the data. During exploration, these frames are compared to new information and discoveries. People decide if these new insights augment their current understanding, refute it, or require a new frame entirely.

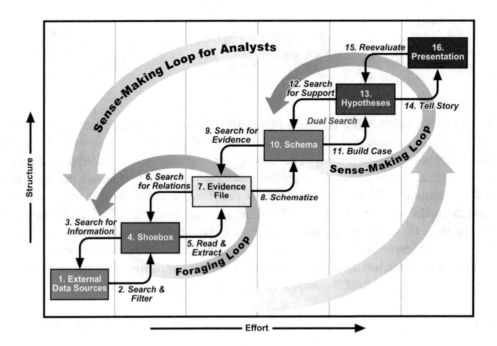

Figure 2.1: The sensemaking loop, modeling the cognitive stages for individual intelligence analysis (adapted from [57]).

There are several other descriptions of sensemaking and analytical reasoning as it pertains to visual data analysis (e.g., [57, 81–83, 93, 94]). In general, many of these works underscore the importance of letting users interactively explore and form mental models of the phenomena represented in the data. Further, sensemaking processes are incremental and continuous. Humans build their cognitive constructs and insights over time and through experiences. This process of mapping a human's understanding of the domain to the analytic models' approximations of the data is central to interactive data analysis.

2.2 THE USE OF ANALYTIC MODELS IN VISUAL ANALYTICS

Visual analytics places an emphasis on the integration of data analytics into the visual data exploration process. That is, often the visualization shown are results of models and computational processes that approximate some phenomena in the data. These models approximate different data characteristics, such as trends, patterns, anomalies, similarity between groups of data items, and others. Thus, the design and implementation of visual analytic applications are largely tai-

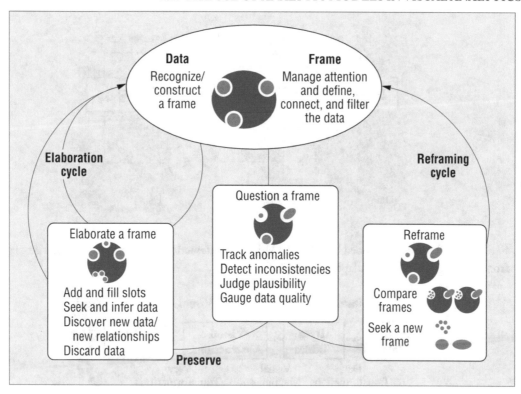

Figure 2.2: Klein's data frame model that depicts the incremental creation of a person's understanding of data (adapted from [83]). The process of creating one's "frame" (or understanding) about data consists of matching one's current frame with new data, and deciding if the current frame should be augmented, broken, or a new one created.

lored around the incorporation of these analytic models into the cognitive processes of people performing data analysis.

Early diagrams and models for information visualization often emphasize the importance of binding visual encodings to data values, selecting visual representations, and other aspects. Each of these are important, yet the goal here is to highlight the added complexity and central role that analytic models play for visual analytics.

For example, Jarke van Wijk's model for visualization shown in Figure 2.3 illustrates the knowledge gain a person may have depending on the choice of visual encodings and visual metaphors used to represent the data [85].

Similarly, a frequently shown illustration of the "visualization pipeline" depicts data transformations, visual mappings, and visualization techniques as the main components of a visualization system. See Figure 2.4.

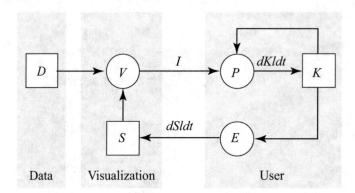

Figure 2.3: A model developed by Jarke van Wijk user interaction as technique to move visualizations from one stage to another (adapted from [85]).

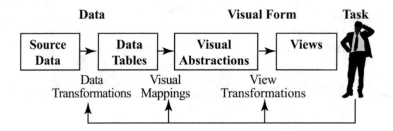

Figure 2.4: The visualization pipeline, showing places where user interaction can tune and adjust parameters to change visualizations and foster exploration (adapted from [33]).

In contrast, diagrams that illustrate the main components of visual analytic systems depict the interplay between analytic models, machine learning techniques, and statistics. For example, Sacha et al. [95] present a model that more specifically outlines the interactions between models, data, and people (see Figure 2.5).

2.3   USER INTERACTION

User interaction is critical to the success of visual analytic applications. It is the mechanism by which people advance beyond being passive observers of visualizations, to becoming active participants in the visual data analysis process. Thus, it is important to understand how to design and implement interactions to enable people to act upon their intuition and domain expertise to properly explore their data.

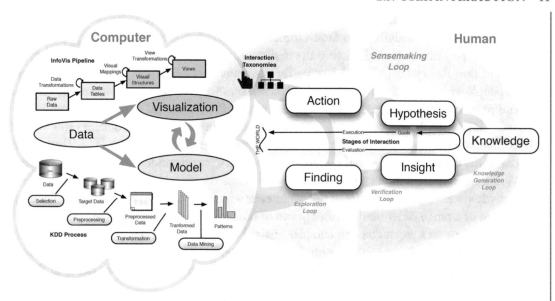

Figure 2.5: A model of visual analytic systems developed by Sacha et al. The model highlights the complexity combining models, user feedback, and interaction into complete applications (adapted from [95]).

One of the challenges for information visualization researchers is to gain a deeper understanding of how people interact with visualizations. There are many potential interactive affordances a visualization designer can add to user interfaces for people to interact with. Yi et al. presented an extensive categorization of such user interactions available in popular exploratory visualization tools [78]. They performed a survey of information visualization applications, enumerating and categorizing the user interaction available to the users. As a result, they found the following user interaction categories:

1. Select

2. Explore

3. Reconfigure

4. Encode

5. Abstract/Elaborate

6. Filter

7. Connect

An additional challenge is understanding how best to weave these user interactions into the people's basic analysis tasks (and ultimately into higher-level sensemaking and reasoning). For example, Elmqvist et al. discuss the design of user interfaces for visualizations through the lens of flow theory [19]). They advocate for designers to consider the holistic analysis processes of people when creating interactive affordances, so as to create the least amount of disturbance, and promote "flow." They refer to this notion of user interaction as "fluid interaction."

Pike et al. have recently called for a "science of interaction" that outlines this connection between user interaction, mental models, visualization, and data analytics [56]. They argue that one of the key advances to visual analytics is through a scientific understanding of user interaction. While the low-level interaction techniques and tasks described above are still present in visual analytic systems, the complexity of the parameters which they control has increased. For example, instead of a range slider used to filter prices of homes for sale, a similar slider may now be used to set bounds on a parameter of a complex analytic model. Thus, understanding how to maintain intuitive and effective graphical controls in an era where system complexity is increasing is an open challenge, worthy of a scientific study. This topic is covered in further detail in the Model Steering section below.

One growing area of research is interested in how to make interactive machine learning more human-centered. Often called "human-centered machine learning" [100–102], these techniques are focused on methods to enhance the user interactions that serve as feedback to the machine learning models. For example, interactive machine learning often utilizes user feedback such as asking users to place labels on data to steer classification models, or giving boolean "thumbs-up" or "thumbs-down" ratings to signify relevance. These interactions are often not directly relevant to the analytic task being performed. Instead, analysts are tasked with high-level goals of understanding, contextualizing, and predicting events from data. A broad set of interactions exist to help analysts explore their data visually, raising the challenge of how we translate these interactions into meaningful feedback to machine learning models.

Even with these select examples, and much more work, user interaction remains an open area of research for visual analytics. The methods and choice of interaction designs are broad, and thus inherently complicated to formally categorize [12]. A more complete survey of user interaction for information visualization and visual analytics can be found in [84].

2.3.1   MODELING USER INTEREST FROM USER INTERACTION

Given the importance and role of user interaction in visual analytic systems, much can be learned from the collection of interactions people perform. Since user interactions are the mechanisms that we place into visual analytic systems that allow users to perform cognitive reasoning tasks, aspects of that reasoning can also be uncovered from the collection of the interactions.

For example, Dou et al. have shown that through logging user interactions in a visualization of financial data, low-level analytical processes can be reconstructed [13, 17]. Their study found that independent researchers can study and manually analyze the interaction logs generated by study participants using a visualization, and recover some of the analytic reasoning of the participants. Most importantly, these results indicate that a detectable connection exists between the low-level user interaction and the analytic process of that user.

Further, there exists work to systematically interpret, approximate, and model analytical reasoning from user interaction. For example, Brown et al. presented the results of modeling characteristics about users, as well as predicting task performance, from user interactions created during a visual search task (i.e., finding Waldo) [96]. Their work showed that basic interactions performed during visual search can be modeled to indicate personality characteristics about the users, as well as predict how well people would perform on the task.

Semantic interaction takes a similar, but slightly different approach. The goal is to systematically analyze the user interactions people perform, and reapply the computed information to steer and adapt the system. Thus, the characteristics about the data and user models are leveraged by the system during the data analysis, for the benefit of the user. One such approach is through model steering, a concept described in the section below.

2.4   MODEL STEERING

Model steering (sometimes referred to as computational steering) consists of the explicit or implicit guidance of analytic models or computation. Computational models are effective at highlighting specific phenomena based on data characteristics in datasets. Each model comes with tradeoffs, designed and developed for specific data types, characteristics, or answers one seeks to gain from performing the computation. This means that there are potential biases or weaknesses inherent in some of these models. For example, Principal Component Analysis, while effective for many dimension-reduction tasks, is susceptible to poor performance when the data includes outliers.

Thus, human-in-the-loop approaches advocate for an incremental, collaborative process to correctly capture and balance the data characteristics that are computationally significant, but also meaningful to the domain [87]. Domain experts can steer these models toward domain-specific characteristics which may not exhibit the most pronounced raw data characteristics. See Figure 2.6.

Models can be steered *explicitly*. This form of steering includes giving users direct control over specific model parameters. These are often exposed to through various graphical control panels and user interfaces. For example, Principal Component Analysis can be steered by giving users direct control over how much each individual dimension of the data should contribute. Figure 2.7 shows iPCA, an example of an interface that allows this form of user control for model steering [39]. Predictive models can be steered by giving people control into the features selection process. For example, Lu et al. showed how this can help people tune prediction models for event detection in social media streams [103]. In general, user interaction as a method for giving people more insight into "black-box" machine learning models is a growing and promising area of research [104].

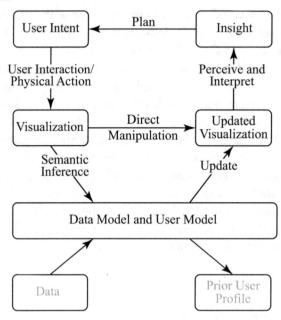

Figure 2.6: A model for human-in-the-loop visual analytics from Endert et al. Their model highlights how user and data models can be inferred via semantic interaction, and integrated into application design (adapted from [87]).

Implicit forms of model steering can be categorized by the actions taken by the user and differentiated from the explicit forms. For example, an implicit form of model steering could be a person marking an email message as spam. The steering that happens as a result of that action typically does not require user input on the specific model parameters. Instead, the user actions are often on the data items, such as labeling or grouping them with other similar items. A more complete survey of implicit and explicit forms of model steering can be found in [97].

2.5 MIXED-INITIATIVE SYSTEM PRINCIPLES

The concept of mixed-initiative systems advocates for a balance between human and machine effort when performing a given task [86]. Systems that exhibit such characteristics must balance both the distribution of tasks performed by either the person or the machine, and also the agreement of the result of these tasks. For example, clustering of similar data items can be performed computationally and manually. However, differences between how people and clustering algorithms group items will likely exist. This raises important user interface and dialog questions as how to resolve such differences. The complete set of design guidelines for mixed-initiative systems can be found in [86].

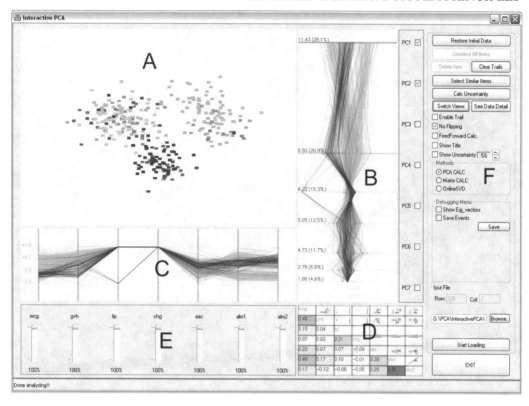

Figure 2.7: A screenshot of iPCA, developed by Jeong et al. [39]. Their application lets people directly modify parameters of PCA to produce updated two-dimensional scatterplots of the high-dimensional data. The interaction is directly on model parameters.

The concept of mixed-initiative systems is integral to many visual analytic applications today. One way that these principles are implemented in such systems is the ability to "take initiative" or perform some actions on behalf of the user. Equally important is to ensure human control or action. Thus, the goal of mixed-initiative systems is to properly balance the action and work between the user and system. The need to explore the appropriateness of this balance has been explored by Endert et al., who argue for "human-is-the-loop" techniques [87]. Such techniques adhere closely to the sensemaking processes described earlier and contrast with more traditional human-in-the-loop guidelines. Specifically, the distinction is made that domain experts should not be used to explicitly tune parameters of collections of analytic models. Instead, their cognitive abilities can be better leveraged by systems that allow users to engage in higher-level analytical reasoning, while systems learn from their actions to steer lower-level computational models.

CHAPTER 3

Spatializations for Sensemaking Using Visual Analytics

This chapter describes how one particular visual metaphor (spatializations) has been used to show the results of analytic models that determine data similarity, but also used by people to organize their data to perform analytic reasoning and sensemaking. This dual-usage makes spatializations a particularly interesting visual medium for semantic interaction. People can reason about data represented in a spatial manner, and two-dimensional scatterplots are commonly computed as a result of dimension reduction, clustering, and other similarity-based algorithms.

Spatializations are visual representations where the primary visual encoding is position between two or more data points. Grounded in cartography, this visual representation emphasizes the concept of distance between data points approximating similarity [66]. That is, points that are closer together are perceived as being more similar than points that are further apart. Holistically, these views often present two-dimensional scatterplots where clusters and groups of similar data points can be seen (when such similarity exists in the data).

The ability for people to make use of spatializations is likely rooted in our innate abilities to make use of our physical and spatial environment everyday. Through the use of our senses, we can place information in locations that occupy meaning to us. The keys you leave on a specific location each day, the sticky note on the refrigerator to remind you to buy milk—they all have meaning given their user-specified location. The use of our physical environment as a means for offloading and externalizing such cues has been frequently studied [88, 89].

Prior work exists that has shown how such spatial layouts of information can be generated by people as well as through computing data positions algorithmically, described in more detail in the sections below.

The current prevalent metaphor that has been explored is a spatialization. This is likely due to the direct link between spatializations and outputs of dimension reduction models. However, additional models can be steered by the bi-directionality of the visual metaphors often used to show their outputs. For example, steering topic models via word clouds, predictive models via spark lines, and other visual manipulations of the metaphors that show the model outputs.

3.1 THE VALUE OF MANUALLY ORGANIZING DATA IN SPATIALIZATIONS

A series of studies have shown how people can make use of space, and spatializations, to accomplish data analysis and sensemaking tasks [5]. Placing and organizing items spatially has been shown to offload aspects of one's working memory into the spatial environment and the spatial constructs contained in it. These studies help inform the research and development of semantic interaction techniques, as one of the guiding principles is to bind model steering functions to user interactions that are native to the domain, data, and visualization technique used. One of the visual metaphors that has been studied is the spatial layout of information on a two-dimensional plane (i.e., a spatialization).

By providing users a workspace in which to manually create spatial representations of the information, users were able to externalize their semantics of the information into the workspace. That is, people can create spatial structures (e.g., clusters, timelines, etc.), and both the structures as well as the locations relative to remaining layout carry meaning to the users with regard to their sensemaking process. Marshall et al. have pointed out that allowing users to create such informal relationships within information is beneficial, as it does not require users to formalize these relationships [47]. Marshall and Rogers [48] found that users prefer to create implicit relationships between pieces of information by positioning related information closer together. They found that the ease, flexibility, and informality associated with creating these relationships spatially were important to users. They refer to this process of incrementally adding more meaning and structure to the spatial layouts as "incremental formalism."

Spatializations allow users to organize and maintain their hypotheses and insight regarding the data in a spatial medium. In large part, this is done through presenting users with a flexible spatial workspace in which they can organize information through creating spatial structures, such as clusters, timelines, stories, etc. In doing so, users externalize their thought processes (as well as their insights) into a spatial layout of the information.

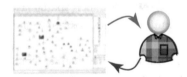

Figure 3.1: A model of interaction with synthesis tools. Users manually create a spatial layout of the information to maintain and organize their insights about the data.

For example, Analyst's Notebook [105] provides users with a spatial workspace where information can be organized, and connections between specific pieces of information (e.g., entities, documents, events, etc.) can be created. Similarly, an application called the nSpace2 Sandbox [75] enables users to create a series of cases (collections of information) which can be organized spatially within the workspace. The benefits of these systems comes less from the computational abilities of analytic models, but from the spatial and interactive affordances given to people to organize their information during sensemaking. From Andrews et al., we learn that users externalize semantic information about a dataset into the layout and organization of documents [5]. The spatial layouts created represent specific meaning about each individual user's analysis. Therefore, the findings from their analysis tasks were present in their spatial layouts. These spatial constructs created by users are complex, and often difficult to accurately characterize computationally [27]. They contain mixed levels of formalism, structure, and meaning, representing fragments of the sensemaking process that the analyst found important to externalize into the spatialization.

3.2   COMPUTATIONALLY GENERATING SPATIALIZATIONS

Spatializations of data can be created by a series of data transformations and computational methods. These include feature selection and extraction, dimension reduction, clustering, and other statistical methods that transform high-dimensional data to lower (often two) dimensions. Several algorithms exist with a similar purpose of mathematically generating two-dimensional layouts from which users can interpret important information about a dataset. In general, algorithms group or organize data based on similarity, which is a function of the features of the dataset. Dimensionality reduction algorithms can provide a 2-d spatial visualization of the clustered data. For example, algorithms like self-organizing maps [66] or generative topographic mapping [41] provide a direct method of visualizing text data spatially, but do not provide explicit cluster membership information. A survey of clustering algorithms can be found in [77] and is outside the scope of this discussion. The primary criteria upon which these models generate layouts are structure extracted from the dataset, such as term frequencies, temporal attributes [2], etc., from textual datasets [74].

There exist several visual analytic applications that have made use of such algorithmic data transformation. This category of applications can be broadly characterized by their ability to pass data through complex statistical models and visualize the computed structure of the dataset for the user to gain insight (Figure 2.1). Thus, the user interaction of these tools is primarily done through directly manipulating graphical controls of parameters that tune and adjust these models. As such, users are required to translate their domain expertise and semantics about the information to determine which (and by how much) to adjust these parameters.

Visual analytic tools such as IN-SPIRE's "Galaxy View" (shown in Figure 2.3) present users with a spatial layout of textual information where similar documents are proximally close to one another [74]. An algorithm creates the layout by embedding the high-dimensional collection of text documents into a two-dimensional view. In these spatializations, the spatial metaphor is one from which users can infer meaning of the documents based on their location. The notion of distance between documents represents how similar the two documents are (i.e., more similar documents are placed closer together). For instance, a cluster of documents represents a group

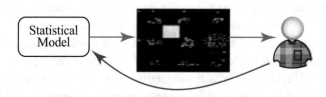

Figure 3.2: A model of interaction with foraging tools. Users interact directly with the statistical model (red), then gain insight through observing the change in the visualization (blue).

of similar documents, and documents placed between two clusters implies those documents are connected to both clusters. These views are beneficial as they allow users to visually gain a quick overview of the information, such as what key themes or groups exist within the dataset. The complex statistical models that compute similarity between documents are based on the structure within the data, such as term or entity frequency. In order to interactively change the view, users are required to directly adjust keyword weights, add or remove documents/keywords, or provide more information on how to parse the documents for keywords/entities upon import.

Alsakran et al. presented a visualization system, STREAMIT, capable of spatially arranging text streams based on keyword similarity [3]. Again, users can interactively explore and adjust the spatial layout through directly changing the weight of keywords that they find important. In addition, STREAMIT allows for users to conduct a temporal investigation of how clusters change over time.

From this related work, the goal is to highlight how user interaction in such tools is designed to be on the model parameters directly, and done through graphical control panels. That is, while the visual representation given to users is spatial, the methods of interaction require users to step outside of that metaphor and interact directly with the parameters of the statistical model using visual controls, toolbars, etc. For example, iPCA (Figure 2.7) discussed earlier is another such example, where controls are on the numerous parameters of the algorithm, such as individual eigenvalues, eigenvectors, and other components of PCA [39].

There has been some work in providing more easy to use interactions for updating statistical models used in visual analytic applications and spatializations. For example, relevance feedback has been used for content-based image retrieval, where users are able to move images toward or away from a single image in order to portray pair-wise similarity or dissimilarity [72]. From there, an image retrieval algorithm determines the features and dimensions shared between the images that the user has determined as being similar. More recently, Ruotsalo et al. created an application with a similar interaction design for information retrieval and search of text documents [90]. Their

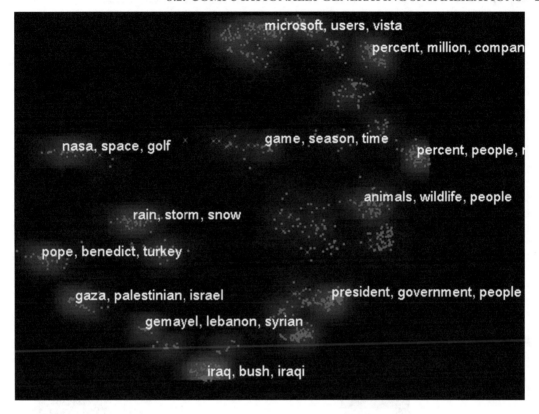

Figure 3.3: The IN-SPIRE Galaxy View showing a spatialization of documents represented as dots. Each cluster of dots represents a group of similar documents.

concept of "interactive intent modeling" allows users to position terms relative to the center of a target. The more central a term is positioned, the more tailored the search results will be to that term.

Also, spatializations of document sets exist that allow users to place "points of interest" directly into the spatial layout. For example, a visual analytic application called VIBE allows users to define multiple points of interest in the spatial layout that correspond to a series of keywords describing a subject matter of interest to the user [52]. Similarly, Dust & Magnet [79] allows users to place a series of "magnets" representing keywords into the space and observe how documents are attracted to or repelled from the locations of these magnets. Through both of these systems, users can interact in the spatial metaphor through these placements of "nodes" representing attributes or dimensions of the data. Additionally, ExPlates by Javed and Elmqvist [107] allows people to create spatial layouts of visualizations states to track their exploration process (including annotating these states for easier recall later).

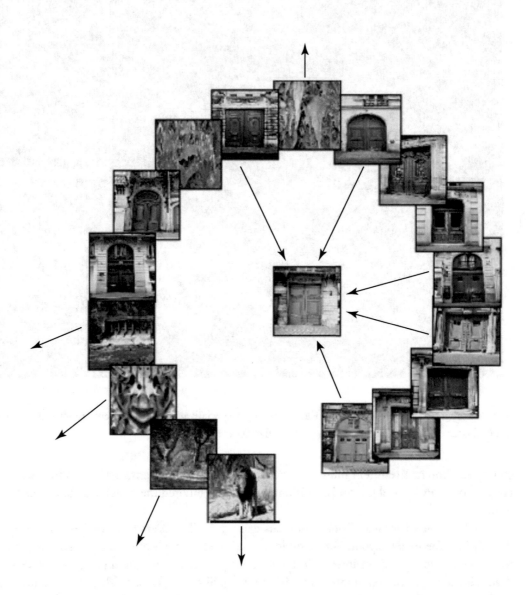

Figure 3.4: A relevance feedback model example. Users move images arranged in a circle toward or away from the image in the middle to signify similarity or dissimilarity. Figure originally published by Stefan Rueger [61]. Used under fair use guidelines.

CHAPTER 4

Semantic Interaction

In the purest sense, semantic interaction is a form of user interaction that couples the exploratory user interaction in visualizations with one or more underlying analytic models. Semantic interaction provides a link from the data in visualizations back to the analytic models that compute them. Often, the interactions occur within a visual metaphor (e.g., a spatialization), where the users interact directly with the visual representation or encoding used to represent the outputs of a model. By interacting directly with this visual output, there exists a tight coupling between the interpretation of the model results, and the user feedback that is incorporated into the computation. Semantic interaction uses the visual metaphor as a bi-direction medium through which people can analytic models can communicate and collaborate about data [105].

To illustrate semantic interaction, the remainder of this chapter will focus on leveraging the bi-directionality of spatializations for text analysis. As described in the previous chapter, spatializations are effective methods to output the results of many analytic models, and also for people to externalize their subjective and expertise-driven opinions about the data (see Figure 4.1). There are additional visual metaphors that may support this form of user interaction. The choice of text analysis also helps articulate the importance of user feedback and model steering. These sensemaking scenarios typically involve reading, analyzing, and understanding a collection of intelligence reports about a potentially threatening situation. Thus, a combination of visualization, analytic models, and human domain-expertise are required to come to a decision.

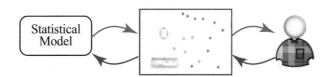

Figure 4.1: A simplified model of semantic interaction that depicts how users do not directly interact with model parameters, but with visual representations that trigger model parameters.

The following intelligence analysis scenario is representative of the strategies and interactions of analysts when performing an intelligence analysis task of text documents in a spatial

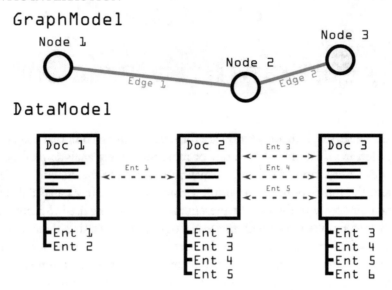

Figure 4.2: Overview of how nodes and edges in ForceSPIRE's force-directed layout are created from documents (Doc) and entities (Ent), respectively.

visualization, as previously found by Andrews et al. [5] and other prior literature that studied intelligence analysis.

During her analysis, an intelligence analyst finds a suspicious and interesting phrase within a document. While reading through the document, she highlights *the phrase "suspicious individuals were spotted at the airport" in order to more easily recall this information later. After she finishes reading the document, she* moves *the document into the bottom right corner of her workspace, in the proximity of other documents related to an event at an airport. To remind herself of her hypothesis, she* annotates *the document with "might be related to Revolution Now terrorist group." Now, with the goal of further examining the events at the "airport," she* searches *for the term, continuing her investigation.*

Via semantic interaction, the system learns from these interactions and provides the analyst with visual feedback. Two documents relevant to the documents in the bottom right corner begin to move closer to that cluster. She quickly reads through these, and notices that one of them seems related, and moves it into the cluster. It informs her that the "Revolution Now group is operating in airports," strengthening her insights. The other document talks about a "terrorist at an airport in Afghanistan." She moves this document away from the cluster, notifying the system that this recommendation was not relevant, as she is not currently investigating activities in Afghanistan. Through incremental learning and user feedback, the layout is co-created from the domain expertise of the user and the computed similarity of the system.

In addition to the three forms of semantic interaction in the scenario, Table 4.1 provides a list of various forms of semantic interaction, including how each can be used within the analytic process of investigating textual information spatially. This list is likely incomplete, but serves as a starting point to introduce how semantic interaction can be integrated into a user's reasoning process.

4.1   DESIGNING FOR SEMANTIC INTERACTION

In order for analysts to interact with information in a spatial metaphor, it must first be created. Following the model of the visualization pipeline [33], this creation calls for a series of data transformations, turning raw data into a spatial layout—much the way many of the visualizations mentioned previously are constructed. Designing for semantic interaction requires a fundamentally different model for how tools integrate user interaction—one that can *capture the interaction*, *interpret the associated analytical reasoning*, and *update the appropriate mathematical parameters*. This chapter describes how these components are performed, using an existing visual analytics prototype, ForceSPIRE [25], as a use case to frame the discussion. ForceSPIRE is a visual analytic tools that uses dimension reduction and entity extraction models to produce a spatial layout of text documents. ForceSPIRE gives users interactions that are native to the domain of the data (e.g., highlighting text, grouping similar documents, etc.) that are interpreted by the system to update and steer the models and support analysis. ForceSPIRE is described in more detail later in the book.

Figure 4.3 illustrates this model, where the spatialization is treated as a medium through which the user can perceive information and gain insight, as well as interact and perform his analysis. Through expanding the pipeline to accommodate for semantic interaction, it is a more appropriate match to the user's sensemaking process. This model emphasizes that user interaction is not intended to be only a method for people to transition from one state of a visualization to the next. Instead, interaction also represents some aspects of analytical reasoning. This form of data can be stored and interpreted systematically, and used to help steer and augment the computational abilities of the system, described below.

4.1.1   CAPTURING THE SEMANTIC INTERACTION

A non-trivial first step in the model is capturing the user interaction. Much research has been done in this area, primarily for the purpose of maintaining process history (e.g., [11, 30, 64, 91], etc.). When considering how to capture interaction, one decision to be made is at what "level" to capture it. For example, GlassBox [15] captures interaction at a rudimentary level (i.e., mouse clicks and key strokes), while Graphical History [34] keeps track of a series of previous visualizations as a user changes the visualization during the exploration of the data.

The level at which user interaction is captured in support of semantic interaction is at a *data level*, as the interactions occur on the data, and within the spatial metaphor. Using the earlier analytic scenario, the interaction being captured would be:

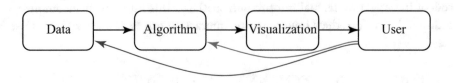

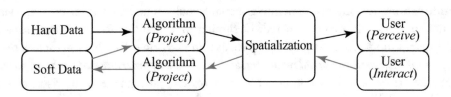

Figure 4.3: (top) The basic version of the "visualization pipeline." Interaction can be performed directly on the algorithm (blue arrow) or the data (red arrow). (bottom) Our modified version of the pipeline for semantic interaction, where the user interacts within the spatial metaphor (purple arrow).

- The highlighted **phrase**
- When the highlighting occurs (**timestamp**)
- The **color** chosen for the highlight
- The **document** in which the highlight occurs
- The new document **location**
- The text of the **annotation**

By capturing (and storing) the interaction history, it is possible to systematically interpret the analytical reasoning of the user. Thus, the user interaction is not only captured, but interpreted and used within the current analytic discourse. The goal is to use the analytic provenance to steer the analytic model and advance the analytic process.

4.1.2   INTERPRETING THE ASSOCIATED ANALYTICAL REASONING

In interpreting the interaction, the system determines the analytical reasoning associated with the interactions and updates the model accordingly. From previous findings [5], categories of user interaction can be associated with specific forms of analytical reasoning (see Table 4.1). It is essentially the model's task to determine *why*, in terms of the data, the interaction occurred, and how that information can be used to augment and adjust the analytic models of the system to help the user's task. The goal is to calculate, based on the data, what information is consistent with the captured interaction. For instance, we can associate text highlighting with adding importance to the text being highlighted.

Table 4.1: Forms of semantic interaction supported in ForceSPIRE. Each interaction corresponds to reasoning of users within the analytic process. Corresponding model updates are performed to steer the model based on the user's reasoning.

Semantic Interaction	Associate Analytic Reasoning	Model Updates
Document Movement	• Similarity/Dissimilarity • Create spatial construct (e.g. cluster, timeline, list, etc.) • Test hypothesis, see how document "fits" in region	• Similarity/Dissimilatiry b/w documents • Up-weight shared entities, down-weight others
Text Highlighting	• Mark importance of phrase (collection of entities) • Augment visual appearance of document for reference	• Up-weight highlighted entities
Pinning Document to Location	• Give semantic meaning to space/layout	• Layout constraint of specific document
Annotation, "Sticky Note"	• Put semantic information in workspace, within document context	• Up-weight entities in note • Append entities to document and model
Level of Visual Detail (Document vs. Icon)	• Change ease of visually referencing information (e.g., full detail = more important = easy to reference)	• (Full Document): "heavier node," increase node's friction • (Icon): "lighter node," less friction
Search Terms	• Expressive search for entity	• Up-weight entities contained in search • Add entities to model

The captured and interpreted interactions are referred to as *soft data*, in comparison to the *hard data* that is extracted from the raw textual information (e.g., term or entity frequency, titles, document length, etc.). We define soft data as the stored result of user interaction as interpreted by the system. In representing interaction as soft data, the algorithm can calculate and reconfigure the spatial layout accordingly. Figure 4.3 illustrates how our approach differs from the traditional visualization pipeline. Soft data can also be thought of as "meta-data," but with the distinction that it is not supplied with the raw data, and it is inferred from the interactions.

There has been previous work in capturing and interpreting reasoning from user interaction. For instance, the study by Dou et al. [17] discussed earlier found that analytical reasoning of users can be recovered from user interaction logs. While semantic interaction has similar goals (i.e., interpreting the analytical reasoning associated with the analysts through an evaluation of the interaction) the semantic interaction model does so through tightly integrating the interaction with the underlying mathematical model. In doing so, the interpretation can be done algorithmically, and the system guided by the user during use. The following section describes how different models can be updated based on the captured and interpreted soft data.

4.1.3   UPDATING THE UNDERLYING MODEL

Visual analytic techniques often make use of one or more models to compute specific data characteristics, and show the results to users visually. For spatialization techniques, such as the one used in ForceSPIRE, variations of dimension reduction models are often used to generate the two-dimensional layouts from higher-dimensional data. Thus, incorporating soft data into the underlying computation can be realized in many ways. Different algorithms have different parameters that can be adjusted to focus on aspects of the data that are interesting to the user. Depending on the algorithm used to compute the spatial layout, the precise parameters being updated will vary.

4.2   EXPLORING THE SEMANTIC INTERACTION DESIGN SPACE

There are many design decisions that factor into the successful implementation of semantic interaction. The following sections discuss some tradeoffs in the broader design space for semantic interaction for designers and developers of visual analytic systems to consider.

4.2.1   THE INTERACTION-FEEDBACK LOOP

In semantic interaction, much emphasis is placed on providing mechanisms for the system to learn the domain expertise of the user through inferring a model weighting scheme. This can come in the form of weighting and parameterizing existing analytic models, but also through augmenting the data currently in the model (e.g., through annotation and other methods for user input). However, equally important is for the system to provide feedback about what analytical reasoning it has inferred (i.e., learned) from the user. The challenge comes in the way of determining the balance between providing this information *explicitly* (e.g., showing dimension weights, asking the user to confirm learned dimensions, etc.) vs. *implicitly* (e.g., updating the spatial layout, etc.).

Providing explicit feedback to the user about the model (and how it was steered) has been previously studied. For example, Liu et al. describe how interactively adjusting the location of points within a spatialization enables the system to learn about dimensions of a dataset that correspond to the user's feedback [45]. Through performing this type of observation-level interaction, the users are given a set of weights that correspond to their newly generated spatialization. As such, this work focuses on explicitly showing the user the dimensions that correspond to their interaction (i.e., the feedback from the system to the users).

Implicit feedback entails providing the feedback of the model through the visualization, rather than explicitly via the weighted dimensions. For example, ForceSPIRE provides an updated spatialization as a result of a semantic interaction. (An entity viewer window also exists, where dimension weighting can be directly adjusted.) Similarly, previous work on observation-level interaction also uses the updated spatialization as a medium for communicating the learned domain knowledge [28].

However, *how can a system support a mixture between these two forms of feedback?* In some situations, people may have the expertise about the model being used, and prefer to see direct representations that show the temporal change. Alternatively, people may not know about (or need to know about) the model, but rather the visual representation that is more native to their mental model of the domain. Also, one can see that as the number of dimensions increase (and become more abstract), explicit feedback may not be effective or meaningful to the users. Further, the results of a user study of ForceSPIRE (where explicit feedback can be obtained by the entity viewer window) shows that users may not prefer, or need, this form of feedback [24]. Similarly, users may require some feedback to gauge what information the system is learning based on their interaction, and given the ability to provide more fine-grained model steering (e.g., steering at the entity level, rather than at the document level).

One possibility is to maintain this feedback within the spatialization. That is, instead of providing a separate view for the explicit feedback, augmenting the spatialization to include this sort of information may be beneficial. For example, ForceSPIRE includes entity underlining within the text of a document to inform users of which keywords are entities in the model. However, this depth of information could be increased, to highlighting words on a color ramp based on their weight. Then, if users find inconsistencies in the entity weighting scheme, adjustments can be made, and the bi-directional learning can continue. Similarly, InterAxis shows the results of the updated model in the new visualization generated, but also directly on the new weights of attributes applied to the model [92].

Similarly, the use cases and studies discussed in earlier sections describe how users perform sensemaking and other analysis activities over an extended period of time. In these cases, analysts incrementally updated the model, changing the weights of parameters with each interaction performed. However, it should be noted that this is not for the purpose of converging on a single, ideal model parameterization. Instead, each instantiation of dimension weights resulted in an individual visualization which helped incrementally build upon the mental models of the analysts.

Therefore, while the interaction-feedback loop in this results in an updated parameterization of the models, it should be noted that it also helps build understanding and insights into the data.

4.2.2   APPROXIMATING AND MODELING USER INTEREST

The adaptation of the analytic models over time are intended to approximate the evolution of user interest and stages of analytical reasoning. When inferring analytical reasoning in the form of a weighting scheme, a choice can be made as to how to translate an interaction to a set of dimensions and weight values. This learning can happen through cleanly inverting the mathematical model, creating a new heuristic by which to learn weights, or some combination of the two.

Examples of directly inverting the mathematical model can be seen in the modified dimension reduction models presented by Endert et al [28]. In these models, great care was taken to ensure that each interaction (i.e., newly positioned data point in the spatialization) corresponds to a learning of weights while maintaining a given amount of stress in the system. The weights are generated through a backward-solving of the weights using the dimension reduction model, and using the user's newly positioned observations. That is, by people giving the demonstrations of where they would like data points positioned in the two-dimensional visualization, we invert the dimension-reduction model to approximate the weighting of dimensions to produce spatializations consistent with the user-defined positions of points.

Using the same dimension reduction model (MDS), Liu et al. [45] performed the same MDS projection step, but inverted the observation-level interaction differently. Instead of directly inverting the projection used in MDS, they use a modified (biased) calculation of the weights for their learning step. This modification can be seen as a combination of a heuristic and a cleanly inverted model for learning.

ForceSPIRE uses a model for learning weights that is more flexible, and thus does not adhere strictly to the inverted model (a force-directed model, in this case). For example, while performing an observation-level interaction in ForceSPIRE results in an emphasis of the similar characteristics between documents moved closer, the remaining weights are equally reduced for normalization of the global weight. As such, there is no direct inversion of the force-directed model, but instead the model is used to calculate the set of characteristics that correspond to the similarity, and the amount of emphasis those characteristics get (i.e., the increase of the weight of those entities) is via a constant.

The decision of inverting a mathematical projection model may be a good fit for systems where the semantic interactions are primarily observation-level interactions. However, other forms of semantic interactions may not lend themselves to directly inverting a projection model. (e.g., highlighting text, performing a search, etc.) For example, highlighting can be automated given the weights of entities. Then, as users manually highlight (or change the highlighting that the system recommended), the system can invert the model used for highlighting to maintain mathematically valid visualizations. The fundamental principles of semantic interaction still apply to these interactions, as they generalize beyond spatializations and observation-level interactions.

4.2.3   CHOICE OF MATHEMATICAL MODEL

To support semantic interaction, the underlying mathematical model must be appropriately tailored to couple with the user interaction. Previous work has shown how to modify popular dimension reduction models, such as Probabilistic Principal Component Analysis (PPCA), Multidimensional Scaling (MDS), and Generative Topographic Mapping (GTM) to allow the direct manipulation of the points within the spatialization (called *Observation-Level Interaction*) [28]. In ForceSPIRE, the focus is on steering a force-directed model for the purpose of analyzing text datasets.

Based on the feedback from evaluating semantic interaction in ForceSPIRE, two aspects of the mathematical model that are important to users are (1) the incremental nature of how the model updates, and (2) the incremental layout update.

Incrementally updating the weight vector, and in turn the spatial layout, is beneficial to users as it closely maps to how users incrementally gather insight about a dataset (i.e., incremental formalism [63]). For example, a minor adjustment to the location of a document within a cluster should not result in complete layout regeneration. Instead, the slight movement of a document within a cluster may reflect the user building a timeline, and thus only local adjustments should be made. For other, broader moves, such as repositioning a document from one cluster to another, the impact on the weight vector may be more severe, resulting in a greater impact on the positioning of other documents. For this capability, we have found that deterministic models, by definition, are less likely to support these minor updates without requiring a broader layout update. Probabilistic models (such as a force-directed model) can achieve a low-stress state given multiple layout variations.

The updating of the spatial layout is equally important, as it provides the opportunity to show the user what has changed from one layout to the other. That is, it provides the user feedback on what the system has learned from their previous semantic interaction. Models that are incremental in nature (where the calculation of the lowest-stress configuration is incrementally obtained) more easily support this concept, as the user can observe the model achieving the state. For example, users can gain insight into both the characteristics of the model, as well as the weighting vector, through observing a force-directed model settling out.

Equally important is **determining the type of task or goal** that a user plans to accomplish within a spatialization. Given the type of interactions shown in Table 4.1, the style of model that each of these drives may differ. For example, defining the shape of a cluster may be better suited to a clustering model, rather than purely a dimension reduction model. Similarly, labeling a cluster (or modifying the label of a cluster) may lend itself to steering of a topic modeling algorithm. Developing such mixed-initiative and mixed-metaphor systems can cover more adequately the spectrum of analytical reasoning that may be associated with various user interactions, as well as develop a more mature design space for semantic interaction.

Table 4.2: Semantic interactions for directly modifying the spatialization can impact both the relative and absolute spatial positions of documents

Relative	Absolute
On Document:	On Document:
• Move toward other document	• Pin to location
• Move away from other document	• Pin to region
• Add to cluster	
• Remove from cluster	
On Cluster:	On Cluster:
• Create/delete cluster	• Define location, region
• Preserve cluster	• Define shape
• Merge/split cluster	

4.2.4   RELATIVE AND ABSOLUTE SPATIAL ADJUSTMENTS

Semantic interaction enables the flexibility to explore hypotheses spatially, such as the relationships within the dataset through direct spatial adjustments. The spatial metaphor provides a rich medium for the interaction, as the spatial locations of documents are meaningful. The meaning is generated via a document's *relative* position to others, as well as its ***absolute*** position in the spatialization. Table 4.1 provides examples of how these two spatial characteristics can be modified. In providing users such flexibility, many challenges become apparent, such as the ones discussed below. With each of these design decisions exist tradeoffs, such as mathematical complexity, completeness of information, simplicity, and user control.

When users interact within the spatialization, the goal of semantic interaction is to infer the characteristics of the dataset upon which to base the change of similarity between documents. To do so, one design decision to make is *how to capture and quantify the interaction*. Depending on the domain, task, and data for which the tool is being designed for, one of the following may provide more desired results.

Boolean similarity updates—updates to the similarity between documents are treated as being either "more similar" or "less similar" than in the previous view. The "amount" of change to the similarity is not taken into direct consideration. Such an implementation may lend itself more to hierarchical or categorical models where similarity is fundamentally based on membership to a topic or cluster. Drucker et al. present an example of using semantic interaction for such sorting and categorization tasks [18]. Their tool enables users to organize documents into folders, where each folder has the ability to suggest other documents that may fit into the folder based on the information currently contained in it. Semantic interaction aspects of this work are that the documents can be added or removed from these folders directly in the spatial "desktop" metaphor.

Scalar similarity updates—updates to the similarity between documents is calculated based on how much the distance between two documents changed from the previous view. That is, decreasing the distance between two documents by 40% will result in a model that decreases the similarity between the two documents. For spatializations that carry a continuum of meaning throughout the space, scalar updates may be more appropriate. However, users may find it more meaningful to conceptualize the update of similarity through Boolean similarity. For example, ForceSPIRE uses Boolean similarity updates while maintaining a continuous spatialization [25].

CHAPTER 5

Applications that Integrate Semantic Interaction

5.1 FORCESPIRE: SEMANTIC INTERACTION FOR SPATIALIZATIONS OF TEXT CORPORA

ForceSPIRE is a visual analytics prototype designed for specific forms of semantic interaction (document movement, text highlighting, search, and annotation) for interactively exploring textual data. The system has a single spatial view (shown in Figure 1.1) where a collection of documents is represented spatially based on similarity (i.e., documents closer together are more similar). Thus, the primary visual metaphor is a spatialization, where users can leverage all the advantages of using space for reasoning described in the earlier sections.

ForceSPIRE is designed for large, high-resolution displays (such as the one shown in Figure 5.1). As semantic interaction emphasizes the importance of visual context in which the interaction takes place (e.g., highlighting text in the context of the document), having the full detail text available in the context of the spatial layout is beneficial over having a single document viewer. Further, the physical presence of these displays creates an environment in which the virtual information (in this case the documents) can occupy persistent physical space. As a result, users are further immersed into the spatial metaphor, as they can point and quickly refer to information based on the physical locations. Also, simply having more resolution and display space is beneficial as it allows data (in this case documents) to be shown in multiple levels of detail.

5.1.1 CONSTRUCTING THE SPATIAL METAPHOR

The spatial layout of the text documents is determined by a modified version of a force-directed graph model [29]. This model functions on the principle of nodes with a mass connected by springs with varying strengths. Thus, each node has attributes of attraction and repulsion: nodes repel other nodes, and two nodes attract each other only when connected by a spring (edge). The optimal layout is then computed by iteratively calculating these forces until the lowest energy state of all the nodes is reached (i.e., the computation settles on a state). A complete description of this algorithm can be found in [29].

ForceSPIRE applies this model to textual information by treating **documents** as **nodes** (an overview is shown in Figure 4.2). The entire textual content of each document is parsed into a collection of entities (i.e., keywords). The number of entities corresponds to the **mass** of each doc-

Figure 5.1: Using ForceSPIRE on a 32 megapixel large, high-resolution display. Photo by Alex Endert, 2012.

ument (heavier nodes do not move as fast as lighter nodes). A *spring* (or edge) represents one or more matching *entities* between two nodes. Therefore, the initial distance metric is a based on co-occurrence of terms between documents. For example, two documents containing the term "airport" will be connected by a spring. The strength of a spring (i.e., how close together it tries to place two nodes) is based on two factors: the number of entities two documents have in common, and the *importance value* associated with each shared entity (initially, importance values are created using a standard tfidf method [42]).

The resulting spatial layout is one where similarity between documents is represented by distance relative to other documents. *Similarity* in this system is defined by the strength of the spring between two documents. A stronger spring (and therefore a larger amount of shared entities) will pull two documents closer together, and thus represent two similar documents.

5.1.2   SEMANTIC INTERACTION IN FORCESPIRE

The semantic interactions in ForceSPIRE are: placing information at specific locations, highlighting, searching, and annotating in order to incrementally change the spatial layout to match their mental model. The primary parameters of the force-directed model that are being updated through this learning model are the importance values of the entities. This section describes the

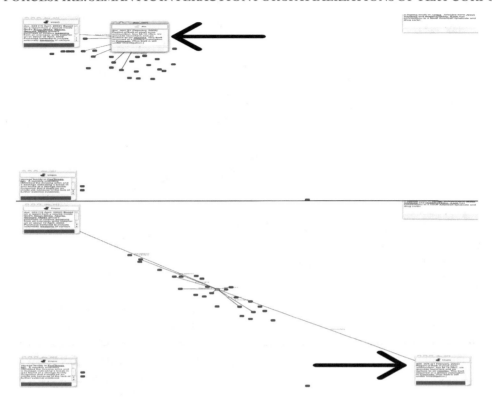

Figure 5.2: Moving the document shown by the arrow, ForceSPIRE adapts the layout accordingly. Documents sharing entities with the document being moved follow.

functionality of these features. The mathematics of how these analytic interactions update the underlying model are presented in Section 5.1.3.

Document Movement. The predominant interaction in a spatial workspace is positioning (and repositioning) documents. In ForceSPIRE, we allow for the following exploratory interaction (i.e., interaction that allows users to explore the structure of the current model, but does not change it). Users are able to interactively explore the information by *dragging* a document within the workspace, *pinning* a document to a particular location (see Figure 5.2), as well as *linking* two documents. When dragging a document, the force-directed system responds by finding the lowest energy state of the remaining documents given the current location of the dragged document. Mathematically, this adds a constraint to the stress function being optimized (in this case the force-directed model). This allows users to explore the relationship of that document in comparison to the remaining documents.

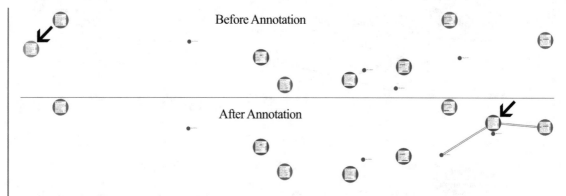

Figure 5.3: The effect of adding an annotation ("these individuals may be related to Revolution Now") to the document shown with an arrow. As a result, the document becomes linked with other documents mentioning the terrorist organization "Revolution Now."

In addition to the exploratory dragging of a document, users have the ability to *pin* a document. By pinning a document, users are able to incrementally add semantic meaning to absolute locations in their workspace. By specifying key documents to user-defined locations, the layout of the remaining documents will adapt to these constraints. Thus, users can explore how documents are positioned based on their similarity (or dissimilarity) to the pinned documents. For instance, if the layout places a document between two pinned documents, it may imply that the particular document holds a link between the two pinned documents, sharing entities that occur in both.

Finally, users can perform an expressive form of this interaction by *linking* two documents, performed by dragging one document onto another pinned document. In doing so, ForceSPIRE calculates the similarity between the documents, and increases the importance value of the entities shared between both documents. As a result, the layout will place more emphasis on the characteristics that make those two documents similar.

Highlighting. When *highlighting a term*, ForceSPIRE creates an entity from the term (if not already one), and the importance value of that term is increased. Similarly, *highlighting a phrase* results in the phrase being first parsed for entities, then increasing the importance value of each of those entities. For example, Figure 5.4 shows the effect of highlighting the terms "Colorado" and "missiles" in the document pointed to with the arrow. As a result, the other documents containing that term are clustered more tightly.

Searching. When coming across a term of particular interest, analysts usually search on that term in order to find other occurrences. In a spatial workspace, this is of particular importance, because the answer to "where the term is also found" is not only given in terms of what documents, but also where in the layout those documents occur. The positions of documents containing the term are shown in context of the entire dataset, from which users can infer the importance of that term (as shown in Figure 5.5).

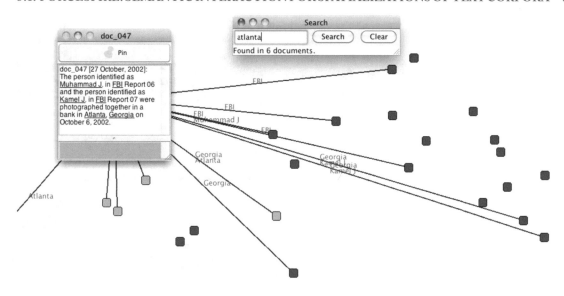

Figure 5.4: The effect of highlighting a phrase containing the entites "Colorado" and "missiles." Documents containing these entities move closer, as the increase in importance value increases the edge strength.

ForceSPIRE first creates an entity from the search term (unless it is already one), then increases the importance value of the search term. Figure 5.5 gives an example of how a search result appears in ForceSPIRE. Searching for the term "Atlanta," documents that contain the term are highlighted green, and links are drawn to show where the resulting documents are in relation to the current document.

Annotation. Annotations (i.e., "sticky notes") are also viewed as a form of semantic interaction, occurring within the analytic process, from which analytic reasoning can be inferred. When a user creates a note regarding a document, that semantic information should be added to the document. For example, if Document A refers to "Revolution Now" (a suspicious terrorist group), and Document B refers to "a group of suspicious individuals," and the user has reason to believe these individuals are related to Revolution Now, adding a note to Document B stating "these individuals may be related to Revolution Now" is one way for the user to add semantic meaning to the document.

ForceSPIRE handles the addition of the note (shown in Figure 5.3) by 1) parsing the note for any currently existing entities, then 2) increasing the importance value of each, and 3) creating any new springs between other documents sharing these entities. In the example in Figure 5.3, edges are created between Document B and Document A (as well as any other documents that mention "Revolution Now"). Additionally, if the note contains any new entities not currently in

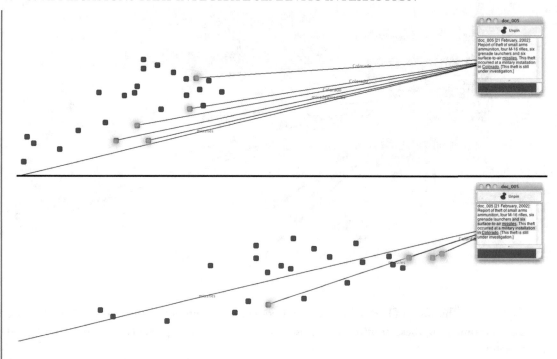

Figure 5.5: Searching for the term "Atlanta," documents containing the term highlight green within the context of the spatial layout. Additionally, the importance value of entity "Atlanta" is increased.

the model, they are created, with the intent that any future entities that may match to that note can be linked at that time. ForceSPIRE also handles cases where notes are edited, with text added or removed from the note, by updating the entities associated with the document, and adjusting the importance values of these entities accordingly.

5.1.3   MODEL UPDATES

A force-directed model is based on a set of nodes connected with springs of different strengths, which pull on each other until the entire graph reaches a lowest-stress state [29]. First, we parse the text of each dataset into T unique entities (i.e., dimensions). Then, a weighting vector is applied to each dimension. As such, each dimension (i.e., entity) in the dataset at the nth iteration has a weight w given by

$$\underline{w}_n = (w_{n1}, \ldots, w_{nT}).$$

Each document consists of a set of entities contained in it, and membership of an entity in a document is defined as

$$x_{it} = \begin{cases} 0, \text{ entity } t \text{ is not in doc } i \\ 1, \text{ entity } t \text{ is in doc } i. \end{cases}$$

Each document is given a mass m. The mass is used to signify how heavy a document is, used in the physics-driven force-directed layout. The mass m of a document i determines the weight of that document in the model. Documents with more weight move more slowly, thus "anchoring" the spatial layout. The mass of a document does not depend on the total number of entities in the document, as each entity only attributes to the mass of the document once. It is described by

$$m_i = \sum_{t=1}^{T} x_{it} w_t.$$

The strength of a spring K (i.e., edge) between two nodes (i.e., documents) can be defined by

$$K_{ij} = \sum_{t=1}^{T} x_{it} x_{jt} w_t.$$

Therefore, a spring with a higher strength will attract the two nodes more closely. In the following sections, we discuss the semantic interactions in ForceSPIRE, and show how the underlying force-directed model is updated.

Document movement   Users can manipulate the spatial layout directly in the spatialization by placing documents in locations based on the user's domain knowledge. These movements (i.e., *Observation-Level Interactions*) can be both *exploratory* and *expressive* [28], differentiated by how they adjust the underlying model. *Exploratory movements* do not change the weighting of keywords (or entities), but use the current weights to determine the position of the remaining documents given the user-defined location of the document being moved. These can be seen as "model constraints," as the user decides the placement of one or more documents, and the model produces the remaining layout based on these static locations. Additional models, and how they can be adapted to support observation-level interaction are discussed in Section 5.2.

With *expressive movements*, users are able to inform the system that the weighting vector should be updated to reflect a change in similarity between two (or more) documents. For example, when placing two documents closer together, the system determines the similarity between those two documents, and increases the weight on the corresponding entities. As a result, a new layout is incrementally generated reflecting the new similarity weighting, where those two documents (as well as others sharing similar entities) are closer together.

In terms of model updates, *expressive movements* are the only ones that change the weight vector of the model. First, a set of entities $\mathcal{M} = \{m_1, \ldots, m_M\}$ co-occur in both documents moved closer are calculated. Then, the weight of each entity, w_m, is described by:

$$w_{n+1,m} = w_{nm} + K,$$

where K is a constant by which the weight, w_m, is updated. This constant can be adjusted based on how quickly and aggressively the user wants the model to adjust. Each of the remaining entities is adjusted equally to ensure that a normalized total energy is maintained in the system. This step of the update is described by,

$$w_{n+1,t} = \max\left(0, w_{nt} - \frac{KR}{(T-R)}\right), \; t \notin \mathcal{M}.$$

Once the updated weight vector is computed, the model updates the spring strengths and document masses, and the layout iterates until settling again.

Users also have the ability to pin documents to specific locations. These documents serve as spatial landmarks, in that they persist at that location, and the force-directed model treats them as layout constraints, organizing the remaining documents around them. Additionally, pinning allows ForceSPIRE to distinguish between exploratory and expressive movements. Dragging a document near a pinned document will briefly color both documents pink to alert the user of the expressive movement (if the user releases the document at this location). Thus, all other movements in the space are exploratory movements.

Text highlighting   Users can highlight text segments directly in the full-detail document views. As a result, the system increases the importance of the terms highlighted, updating the underlying mathematical model, and ultimately the layout. The phrases highlighted are also parsed for entities using a more aggressive entity extraction algorithm, so as to add entities to the model that may have been initially missed. Users also have the ability to select specific entities to add, delete, or modify by using the Entity Viewer.

When highlighting a phrase of text within a document, ForceSPIRE creates a set of entities contained in the highlight. By using a more aggressive form of entity extraction on the highlighted phrase, some of the entities found in the highlight may be newly discovered. Those entities are added to the model before updating the weighting vector. Suppose the user highlights a phrase, composed of a set of entities $\mathcal{H} = \{h_1, \ldots, h_H\}$. The updated weight, w_h, of an entity h based on a highlight, is described by:

$$w_{n+1,h} = w_{nh} + K, h \in \mathcal{H},$$

where K is a constant by which the weight of the highlighted entity, w_h, is updated. Each of the remaining entities is adjusted equally to ensure that a normalized total energy is maintained in the system. This step of the update is described by,

$$w_{n+1,t} = \max\left(0, w_{nt} - \frac{KH}{(T-H)}\right), \; t \notin \mathcal{H}.$$

Once the updated weight vector is computed, the model updates the spring strengths and document masses, and the layout iterates until settling again.

Search   Search allows users to perform a standard text search within the dataset. As a result, documents containing the search term will be highlighted, and an edge between the search box and those documents will be created (multiple search boxes can exist). The model is updated by increasing the weight of the entity searched for (and creating a new entity for the search term if it does not already exist).

Searching for a term creates a new node in the graph consisting of only the entity contained in the search. This node can be pinned to define an absolute location in the spatialization given the search term, or can be positioned by the model. The updated weight, w_s, of a search on entity s, is described by:

$$w_{n+1,s} = w_{ns} + K,$$

where K is a constant by which the weight of highlighted entity, w_s, is updated. Each of the remaining entities is adjusted equally to ensure that a normalized total energy is maintained in the system. This step of the update is described by,

$$w_{n+1,t} = \max\left(0, w_{nt} - \frac{K}{(T-1)}\right), \ t \neq s.$$

Once the updated weight vector is computed, the model updates the spring strengths and document masses, and the layout iterates until settling again.

Annotation   An annotation can be individually added to each document. Through annotating a document, users can add "meta-information" to the document based on their domain expertise. For example, adding a note "relates to the events in Chicago" results in parsing the note for entities (i.e., "Chicago"), and adding them to the document, which creates edges to other documents containing "Chicago."

Each of these semantic interactions creates **soft data**, a quantitative representation of captured user interaction within the context of the dataset. Figure 4.3 models how soft data is collected (i.e., captured and interpreted interactions within the context of the dataset), as well as how it is combined with the *hard data* to produce the spatial layout. As a result, the soft data steers the underlying force-directed model. Also, soft data serves as a log of the entity weighting throughout the user's analytic process, and can be examined at any time to gain insight about their process.

The entities included in an annotation $A = \{a_1, \ldots, a_A\}$ (which can include newly identified entities not previously in the dataset) update both the mass of a node as well as the weighting vector as follows. The updated weight, w_a, of an entity a contained in an annotation is described by:

$$w_{n+1,a} = w_{na} + K, \ a \in A$$

where K is a constant by which the weight w_a is updated. The constant can be adjusted by the user or the developer. Our experience working with this particular constant is that it depends on the dataset size, complexity, and personal preference of the user. For our studies, we typically set it to a low, conservative value, which results in slower model updates. Each of the remaining entities is adjusted equally to ensure that a normalized total energy is maintained in the system. This step of the update is described by,

$$w_{n+1,t} = \max\left(0, w_{nt} - \frac{KA}{(T-A)}\right), \ t \notin A.$$

Once the updated weight vector is computed, the model updates the spring strengths and document masses, and the layout iterates until settling again.

Undo   When *undoing* an interaction using the standard "Control+Z" keyboard shortcut, a linear history of the interactions will be reversed, and the importance values of affected entities will be returned to their prior values (as well as document masses). As for the locations of the documents, the reverted importance values and document masses will be responsible for updating the layout. However, this does not guarantee that the layout will return to the exact previous view, and the user may find it necessary to perform small adjustments.

The model updates used in ForceSPIRE serve as an initial approach at how to couple semantic interactions with model updates. Other, more complex methods may exist, and we encourage further research in this area. Sensemaking is a complex exploratory process. As such, semantic interaction can enable analysts to explore their hypothesis *in situ*, while the provenance of their insights is captured and stored. An open area of research is what analyzing the soft data might reveal about the analytic process. For instance, if the importance values of entities converge on a small number of entities, specific biases might be revealed. Similarly, instances during the analysis when new hypotheses are being explored may be indicated by diverging importance values.

5.2   SEMANTIC INTERACTION FOR DIMENSION REDUCTION MODELS

Semantic interaction has been applied to specific dimension reduction models. For this specific subset of techniques, the term "observation-level interaction" is often used. In general, observation-level interaction refers to interactions, occurring within a spatialization, that enable users to interact directly with data points (i.e., observations). A spatialization in this context refers to a two-dimensional layout calculated from high-dimensional data where the metaphor of relative spatial proximity represents similarity between documents. That is, data points placed closer together are more similar. Observation-level interactions are therefore tightly coupled with the underlying mathematical models creating the layout, thus allowing the models to update parameters based on the interaction occurring (realizing the semantic interaction design guidelines). While numerous forms of interaction may exhibit these characteristics (e.g., moving clusters of documents, marking regions of interest within the spatialization, etc.), this section will focus on one—the movement of observations or data items.

There are two primary forms of observation-level interaction: *exploratory* or *expressive*, based on the particular analytical reasoning associated with the interaction, and also how the system responds. During an exploratory interaction, users utilize the algorithm to explore the data and the space. For example, through dragging one observation within the layout, users gain insight into the structure of the data by observing how other data react given the algorithm. While an observation is dragged through the layout, the algorithm adjusts the layout of the remaining data according to how the algorithm computes similarity. Thus, when the observation is dragged toward a cluster of data, similar data points attract, while dissimilar ones repel. Additional information such as a list of similar and dissimilar parameters can also be displayed. Through this process, users learn about a single observation, and how it relates to the other observations in the dataset.

An expressive interaction is different, in that it allows users to "tell" the model that the criteria (i.e., the parameters, weights) used for calculating the similarity need to be adjusted globally. This form of interaction performs the model steering aspects of semantic interaction. For example, as a user reads two documents, she denotes they are similar by dragging them close together. If this were exploratory, the two documents would repel again. However, in an expressive form of this interaction, it is the responsibility of the underlying mathematical model to calculate and determine why these documents are similar, and update the model generating the spatial layout accordingly. Using the methods below, we illustrate how both expressive and exploratory forms of observation-level interaction are enabled through modifications made to three common statistical methods (PPCA, MDS, and GTM).

The mathematical formulations and use cases in this section are originally presented by Endert et al. [28].

5.2.1   PROBABILISTIC PRINCIPAL COMPONENT ANALYSIS (PPCA)

Principal Component Analysis (PCA) [40, 54, 71] is a common, deterministic method used to summarize data in a reduced dimensional form. The summary is a projection of a high-dimensional dataset in the directions with the largest variance. When only two directions are chosen, PCA may produce a spatial representation or map of the data that is easy to visualize. One problem with PCA is that important structures (e.g., clusters) in data may not correlate with variance. Thus, PCA spatializations may mask information in the data that analysts may find useful.

Probabilistic PCA [70] is, simply, a probabilistic form of PCA. This means that PPCA is not a deterministic algorithm, but a statistical modeling approach (specifically, a factor modeling approach) that *estimates* low-dimensional representations of high-dimensional data. Let $\mathbf{d} = [d_1, \ldots, d_n]$ represents a $p \times n$ high-dimensional data matrix, where n represents the number of observations, p represents the number of columns, and d_i (for $i \in \{1, \ldots, n\}$) represents a $p \times 1$ vector for observation i. Also, let $\mathbf{r} = [r_1, \ldots, r_n]$ represent a low-dimensional analogy of $\mathbf{d}$, such that $\mathbf{r}$ is $q \times n$ and $q < p$. For our purposes, we set $q = 2$. PPCA models $\mathbf{d}$ as a function of $\mathbf{r}$,

$$d_i \,|\, W, r_i, \mu, \sigma^2 = No(W r_i + \mu, I_p \sigma^2)$$

where, $No(.,.)$ represents the Multivariate Normal Distribution; μ represents a $p \times 1$ mean-vector of $\mathbf{d}$; $\mathbf{W}$ is a $p \times q$ transformation matrix known as the factor loadings of $\mathbf{d}$; $\mathbf{I}_p$ is a $p \times p$ identity matrix; and σ^2 represents the variance of each dimension in $\mathbf{d}$. By convention, PPCA models each r_i with a Multivariate Normal Distribution centered at zero and with unit variance: $r_i \sim$

$No(\mathbf{0}_2, \mathbf{I}_2)$. In turn, the conditional posterior distribution for r_i is $No(\eta, \Sigma_r)$, where

$$\eta = (W'W + I_2 \sigma^2)^{-1} W' (d_i - \mu)$$

$$\Sigma_r = (W'W\sigma^{-2} + I_2\sigma^2)^{-1}. \tag{5.1}$$

A spatialization of data $\mathbf{d}$ that relies on PPCA plots the posterior expectation η. Similar to PCA, the coordinates η rely on the variability observed in $\mathbf{d}$. To see this, let Σ_d represent the marginal variance of d_i, ($\Sigma_d = V[d_i|\mathbf{W}, \mu, \sigma^2]$). Since $\Sigma_d = W'W + I_2\sigma^2$, we can rewrite η as $\eta = \Sigma_d^{-1} \mathbf{W}(d_i - \mu)$ which shows that the relationship between Σ_d and η is well defined.

The final step in PPCA is to estimate the model parameters, $\{\mathbf{W}, \mu, \sigma^2, \Sigma_d\}$. We take a Bayesian approach. We specify either reference or flat priors for each unknown (as suggested by [70]) and use Maximum A Posteriori (MAP) estimators to assess (and plot) η. For example, when we assign $\pi(\Sigma_d) \propto 1$, the posterior distribution for Σ_d is an Inverse Wishart (IW) distribution,

$$\pi(\Sigma_d | d) \propto IW(nS_d, p, n - p - 1) \tag{5.2}$$

where $\mathbf{S}_d$ represents the empirical variance of $\mathbf{d}$. The MAP estimate of Σ_d is $\mathbf{S}_d$.

User-guided PPCA

To enable analysts to guide PPCA via the data visualization, we take advantage of the relationship between Σ_d and η. Namely, changes in Σ_d will effect η, and changes in η will effect Σ_d, when we invert Equation (5.1).

After obtaining an initial PPCA display, the user adjusts the locations of two observations; i.e., adjusts two columns in η. If the two observations are moved close to one another, the analyst is conveying that in her mental map, the observations are more similar than what they appear in the display; and, if the observations are dragged apart, the analyst is conveying that the observations differ more than what they appear.

The challenge in BaVA is to parameterize the cognitive feedback and update the visualization [38]. First, we determine the dimensions of the data $\mathbf{d}$ for which the adjusted observations are similar and different. Second, we transform the adjustments to η into a hypothetical $p \times p$ variance matrix. We denote this matrix by $f^{(p)}$, as it is a quantified version of $f^{(c)}$. In $f^{(p)}$, the dimensions for which the adjusted observations are similar have small variances and the dimensions for which adjusted observations differ have large variances. Third, we consider the hypothetical variance $f^{(p)}$ to be a realization of a Wishart distribution that has an expectation equal to Σ_d. Finally, we apply Bayesian sequential updating [67, 73] to adjust Equation (5.2) by the parametric feedback $f^{(p)}$,

$$\pi\left(\Sigma_d | d, f^{(p)}\right) = IW\left(pS_d + v f^{(p)}, p, n + v - p - 1\right)$$

where v is solved from a specification $\kappa(\kappa \in [0, 1])$ made by the analyst that states how much weight to place on the feedback relative to the data. Namely, the updated MAP estimate for Σ_d

is a weighted average of the empirical variance $\mathbf{S}_d$ and feedback $f^{(p)}$

$$MAP(\Sigma_d) = \frac{\upsilon}{\upsilon + n} f^{(p)} + \frac{n}{\upsilon + n} S_d$$

thus $\upsilon = n\kappa/(1 - \kappa)$. Now, the PPCA projection of the data $\mathbf{d}$ that is based on $MAP(\boldsymbol{\Sigma}_d)$ will portray both information in the data and expert feedback.

Usage Scenario

A sensitive issue for taxpayers, parents, children, educators, and policy makers is whether an increase in money devoted to education will increase education quality. Money provides a means to buy modern textbooks, employ experienced teachers, and provide a variety of classes and/or extra curricular activities. Although, do the students who benefit from these high-priced resources actually improve academically?

In 1999, Dr. Deborah Guber compiled a dataset for pedagogical purposes to address this question [31]. Based on the following variables, the dataset summarizes the academic success, educational expenses, and other related variables in 1997 for each U.S. state: the average exam score on the Standard Aptitude Test (SAT); the average expenditure per pupil (EXP); the average number of faculty per pupil (FAC); the average salary for teachers (SAL); and the percentage of students taking the SAT (PER). To increase the complexity of the dataset slightly, we added two variables from the National Center for Education Statistics (www.http:nces.ed.gov): the number of high school graduates (HSG) and the average household income (INC). We hypothesize that states that spend more on education will cluster with states with high SAT averages.

To assess the hypothesis and explore the data, we implement the BaVA process using PPCA. Figure 5.6a displays our initial view of the data. Notice that the visualization does not present any structure in the data. For analysts in the field of education, notice that two states with different expectations for SAT scores are displayed close to one another. Thus, we select the appropriate observations and drag them apart as an expressive interaction to obtain an updated view that is displayed in Figure 5.6b. There are two clusters in Figure 5.6b. These clusters correspond with SAT scores above and below the national median.

Based on our hypothesis, we suspect that the clustering structure in SAT relates to EXP. However, when we re-plot Figure 5.6b and label the upper and lower EXP 50% quantiles in Figure 5.6c, EXP does not explain the clusters. Thus, we used a bi-plot to identify which variables explain the structure we see in Figure 5.6c. When we mark the observations above and below the empirical PER median in Figure 5.6d, we see that PER and SAT clearly relate to the formation of clusters in the dataset. Thus, further analyses of SAT and EXP must control for PER.

5.2.2 MULTI-DIMENSIONAL SCALING (MDS)

We extend our framework to another deterministic method, which forms the basis for a large number of visualization techniques: Multi-Dimensional Scaling (MDS).

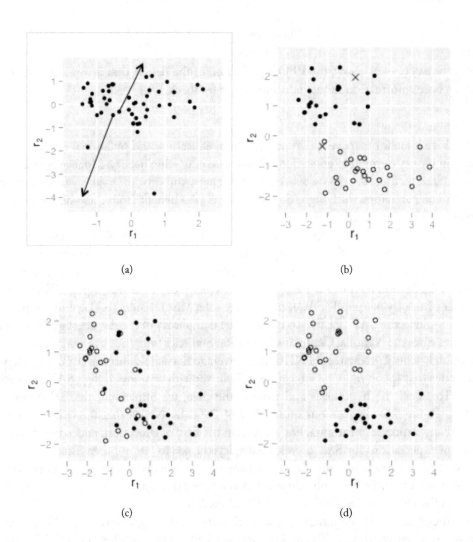

(a)

(b)

(c)

(d)

Figure 5.6: After injecting expert feedback into a), we obtain Figures b)–c). For frame of reference, we marked the two points moved to inject feedback by 'x' in Figure b). The configuration of points in each graph are identical, but the observations are labeled differently. In Figure b), symbols '•' and 'o' mark the upper and lower 50% quantiles for SAT scores, respectively; in Figure c), symbols '•' and 'o' mark the upper and lower 50% quantiles for EXP scores, respectively; and in Figure d), symbols '•' and 'o' mark the upper and lower 50% quantiles for the percentage of students taking the SAT (PER), respectively. Notice the clusters in each graph correspond with SAT and PER, but not EXP.

All complex data visualizations are based on high-dimensional datasets, which contain features corresponding to dimensions and the relative importance of such features through a set of weights (w_i). Classically weighted multidimensional scaling deals with mapping a high dimensional dataset $\mathbf{d} = [d_1, \dots, d_n]$ into a low dimensional (in our case two-dimensional) space $\mathbf{r}$, by preserving pairwise distances between observations in the low dimensional representation. Let $\mathbf{w}$ represent the p-vector of feature weights: $\mathbf{w} = \{w_1, \dots, w_p\}$. Given a set of feature weights, the low dimensional spatial coordinates are found by solving:

$$\min_{r_1, \dots, r_n} \sum_{i < j \leq n} \left| \|r_i - r_j\| - \delta_{i,j}^{(w)} \right|,$$

where

$$\delta_{i,j}^{(w)} = \sum_{k=1}^{p} w_k \, \text{dist}(d_{ik}, d_{jk})$$

such that $\sum_k w_k = 1 \sum_d w_d = 1$. dist() represents any distance function for measuring individual features in the high dimensional space. Because it is not possible to estimate weights and the set $\mathbf{r}$ simultaneously, we provide a uniform weighting of the space $w_i = 1/p$ for our first iteration.

User-guided MDS

Once a visualization is generated, the user may either agree with the display and learn from certain aspects of the visualization, or disagree, based on their domain expertise. Hence, the user may wish to interact and rearrange a few of the observations in the visualization. Given a spatial interaction in the form of adjusting the relative position of a set of points, we compute a set of feature weights, which are consistent with both the user's adjustment and the underlying mathematical model. These are computed by inverting the optimization, by fixing the locations of the adjusted points and finding an optimal set of weights, which are consistent with the visualization. Explicitly, we solve for $\mathbf{w}$ such that

$$\min_{w_1, \dots, w_p} \sum_{\tilde{r}_i, \tilde{r}_j \in M} \left| \|\tilde{r}_i - \tilde{r}_j\| - \delta_{i,j}^{(w)} \right|$$

where $\sum_k w_k = 1 \sum_d w_d = 1$, and M the set of adjusted observations (r_i, r_j). It should be noted that computing the new weights is extremely fast, and is then followed by a full MDS step. Thus, the entire generation of a new view can be performed in real time, depending on the size of the dataset and the specific hardware used.

Usage Scenario

Consider for example a visualization produced by a standard MDS technique. In this example we focus on the 1990 census dataset [9] under a Classical Metric Scaling (CMS) [62], using a Hamming distance (due to the categorical nature of the dataset) for measuring features in the high dimensional space. Figure 5.7 illustrates results obtained under a Classical Metric Scaling (CMS).

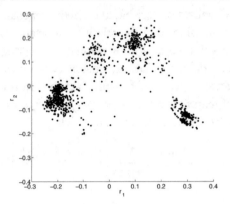

Figure 5.7: Visualization of the 1990 census dataset using classical MDS.

Given this visualization, a user may distinguish three to five main clusters, and inquire what they mean. We see two major ways a user can interact with the visualization, in order to explore the space and learn about the underlying dataset. The first of these is by highlighting a subset of the data, based on some question the user seeks to answer, and then rearranging the visualization based on inconsistencies with their mental model (expressive interactions).

The second approach is to hone in on visual structure, and move points in the visual space in order to learn what the structure relates to in terms of the feature space (Figure 5.8). Both of these interactions are nearly identical, however the motivation for the interactions will differ. We will illustrate both types of visual reasoning through an example based on the 1990 census dataset.

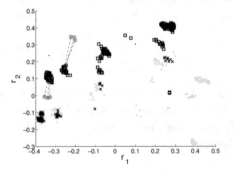

Figure 5.8: A user performing observation-level interaction to learn what distinguishes two clusters.

Figure 5.8 shows how the user might perform an observation-level interaction in order to learn what explains the clustering structure between the working/low-income groups. To suggest the clusters could be moved farther away from each other than they appear in the current visualization, the system reports back the weights, which explains the differences in the groups. For this example, the user learns that one of these clusters contains individuals that have a reliable mode of transportation to work (93% explained). The visualization could be updated based on this information, or the user could simply document this fact and proceed by explaining other areas of the spatialization. As always, there are an endless number of possibilities for learning about a high dimensional dataset via visual expression/exploration. Another example of an exploratory interaction with MDS is demonstrated by Buja et al. in which users can constrain observations to specific spatial locations [10].

The user may wish to interact expressively and identify points in the space that pertain to high- and low-income groups. The user highlights individuals with incomes below 15K and over 60K, as shown by ■ and × in leftmost panel of Figure 5.9, respectively. Because of the close proximity of the highlighted groups in the main clusters, the user drags (denoted by ⊗) a few representative low- and high-income individuals into sets of groups in each of the three main sub-clusters. The system reports back a set of weights, which explain how much a particular feature explains the arrangement of points suggested by the user. High weights relate to important features, while low weights suggest their corresponding features do not relate to the user's visual rearrangement. For our example, we learn not only that income level (29%), but also their means of transportation to work (20%), whether or not they worked the full year (25%), and their level of education (10%) are related to the user's repositioning of points. Given this information, the system updates the visualization, as shown in center panel of Figure 5.8. We notice that in the resulting visualization, the income groups are clearly separated. The resulting visualization displays a much richer spatialization than simply showing clusters relating to the income groups. For example, we highlight individuals who actually worked in the right most panel of Figure 5.8, and notice these individuals are shown in distinct sub-clusters. Two of the four clusters in which individuals work pertain to low-income groups, and the other two pertain to high-income groups (as illustrated by the ■ and × symbols).

5.2.3   GENERATIVE TOPOGRAPHIC MAPPING (GTM)

Introduced by Bishop et al. [14], Generative Topographic Mapping (GTM) is a nonlinear latent variable modeling approach for high-dimensional data clustering and visualization. It is considered to be a probabilistic alternative for both the Self-Organizing Map (SOM) algorithm [43] and Nonlinear PCA. Similar to PPCA, GTM estimates a latent variable $\mathbf{r} = [r_1, \ldots, r_n]$ ($q \times n$ matrix) that is a low-dimensional representation of high-dimensional data $\mathbf{d} = [d_1, \ldots, d_n]$ ($p \times n$ matrix such that $p > q$). However, unlike PPCA, the q-dimensional coordinates $\mathbf{r}$ in GTM map nonlinearly to a complex manifold $\mathbf{m} = [m_1, \ldots, m_n]$ that is embedded in the high-dimensional space. This manifold, ideally, characterizes important structure in data $\mathbf{d}$ and represents geometrically the expected value for $\mathbf{d}$ in the Gaussian model,

$$d_i \ : \ N(W\Phi(r_i), I_p\beta^{-1}). \tag{5.3}$$

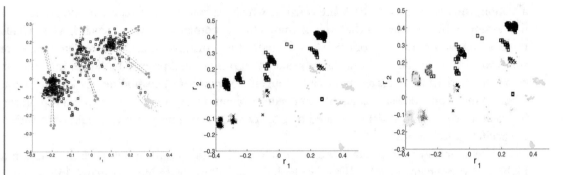

Figure 5.9: A sequence of visualizations derived through observation-level interaction with a modified MDS method. (Left) The user moves a set of points into new locations, communicating his intuition that there may be additional structure within each cluster. (Middle) The updated visualization showing new clusters. (Right) Highlighting showing the separation of income groups in the updated visualization.

To estimate a coordinate m_i, GTM takes a weighted average of J radial basis functions $\{\Phi_1(), \ldots, \Phi_J()\}$ ($\Phi_j()$ represents a radially symmetric Gaussian kernel) given r_i and parameters there in,

$$m_i = W\Phi(r_i) \tag{5.4}$$

$$\Phi_j(r_i) = \exp\left(-\frac{\|r_i - \mu_j\|^2}{2\sigma^2}\right), \tag{5.5}$$

where $\mathbf{W}$ is a $p \times J$ transformation matrix; $\boldsymbol{\Phi}(r_i)$ is a $J \times 1$ vector such that $\boldsymbol{\Phi}(r_i) = [\Phi_1(r_i), \Phi_2(r_i) \ldots, \Phi_J(r_i)]'$; and μ_j is a $q \times 1$ vector that centers the basis functions. The center coordinates $\mu = [\mu_1, \ldots, \mu_J]$ cover the q-dimensional latent space uniformly. Model parameters are estimated using the EM algorithm [16].

One advantage of GTM is that, by construction, it lacks sensitivity to outliers. For tractability, the coordinates of each r_i are limited *a priori* to a finite set g of K possibilities, $r_i \in g = \{g_1, \ldots, g_K\}$ that covers the q-dimensional latent space uniformly. To decide which value for r_i generates d_i, GTM estimates the posterior probability, i.e., *responsibility*, that $r_i = g_k$. Given a prior probability that $r_i = g_k$ is $1/K$ for all $k \in \{1, \ldots, K\}$, let R_{ik} represent the posterior responsibility that latent variable r_i generates d_i, when $r_i = g_k$,

$$R_{ik} = \frac{\pi(d_i \mid r_i = g_k, W, \Phi())}{\sum_{l=1}^{K} \pi(d_i \mid r_i = g_l, W, \Phi())}. \tag{5.6}$$

In turn, GTM plots the posterior mode, expectation, or any quantile of r_i given specifications g and estimates for $\{R_{i1}, \ldots R_{iK}\}$.

User-guided GTM

GTM is a complex modeling approach that relies on many tunable parameters that are hard to interpret. User Guided GTM (ugGTM) will allow analysts to both take advantage of the benefits of GTM and guide the complicated GTM parameterization. Specifically, analysts may label, i.e., *tag* clusters, *tag* regions of the visualization space, and query differences in documents.

Here, we illustrate ugGTM within the context of an example. We have a collection of 54 abstracts from proposals funded by the National Institutes for Health (NIH). After standard preprocessing, we apply a ranking system that we will call an Importance Index (ImpI), which is based on the Gini coefficient. ImpI considers both the frequency and uniqueness of words that are shared across documents and assigns a metric between 0 and 1. Entities that occur equally frequently in all the documents have ImpI=0, and entities that occur in only one document have ImpI=1. We selected the 1,000 entities with the highest ImpI. One advantage of ImpI is that we can measure document similarity using Euclidean distance between proposals. Pairs of documents with small Euclidean distances have comparable terms with similar frequency; and pairs of documents with large Euclidean distances have few, if any, words in common.

We apply GTM for J=16 and K=400 to obtain an initial display of the proposals, shown in Figure 5.10. Notice four clusters appear in Figure 5.10 that we labeled A, B, C, and D.

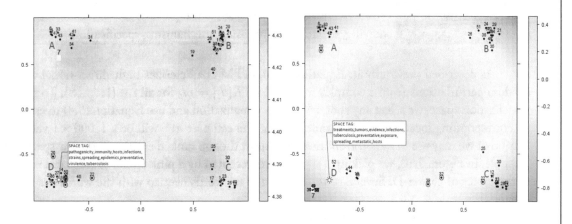

Figure 5.10: GTM display of the NIH abstracts. Black dots mark documents and labeled by their document ID. (Right) shows the updated view after performing an observation-level interaction on the shown points.

Tagging the Clusters and the Space. To understand the meaning of the clusters, we determine the words that both overlap the least within each cluster and have the highest ImpI's. Specifically, we apply k-means [46] to the low-dimensional data coordinates to determine cluster memberships. For each cluster we sum the ImpI vectors across the documents and rank the entities based on the ImpI sum. Entities ranked highest are those that (1) have importance in the corpus (as determined by the ImpI) and (2) have occurred most frequently. Given top rankings from each cluster, we delete those shared by all four clusters. Table 5.1 lists the unique keywords that describe each cluster. Group A represents proposals that include brain-related cancer studies and their clinical applications. Group B represents proposals related to human neural systems. Group C represents proposals that address genomic and transcriptomic research problems. Group D represents proposals about infectious diseases, such as tuberculosis, and immunity.

Table 5.1: Cluster tags (top 10 keywords) for NIH abstract groups

Group A	tumors, brains, stem, treatments, patients, generations, drugs, ordering, controlling, therapeutics
Group B	stem, neuronal, brains, proteins, deliveries, regulations, neural, patients, differentiation, expression, treatments
Group C	stem, genetically, regulations, drugs, structurally, proteins, genomics, epigenetics, RNAs, complexities
Group D	infections, treatments, tuberculosis, expression, patients, drugs, strains, resistance, vaccination, immunity
Shared by All Groups	cells, functionalization, diseases, developments, genes, cancerous, studying, researchers, proposing, mechanisms, specification

As described previously in Equation (5.3), GTM characterizes high-dimensional data as random perturbations from a complex manifold $\mathbf{m}$; $E[d_i] = m_i$ for all $i \in [1, \ldots, n]$. To tag the visualization space, we select any spot, r^+, in the visualization and use Equation (5.4) to estimate its corresponding location on the manifold, m^+. The estimate m^+ will be a $1,000 \times 1$ vector of ImpI that we may use to rank the entities. We report the top ranked entities to tag the space. For example, in Figure 5.10, we pick up a spot r^+ (represented by a pink circle) that locates roughly at the center of cluster D. Several of the tagged top keywords overlap with the words describing cluster D.

Document-based Query and Cluster Reorganization. It is common for users to assess documents by searching for keywords. However, keyword searching may be a tedious task and fail to reveal document clusters of interest. For example, keyword searches may identify documents with similar keywords, but that are used in different contexts; miss documents that contain combinations of the keywords; or prioritize words that have little relative importance for the user. In response to the challenges of keyword searching, many analysts rely on document matching. For document matching, entire documents can be used to identify which of the remaining documents in the corpus are most similar (to the chosen document). Hence, such a matching algorithm is a document-based query of a corpus.

In our ugGTM, users may query documents in the corpus by dragging a document of interest directly in the visualization and watching how the remaining documents respond; e.g., similar documents will follow the document being dragged and dissimilar documents will repel. The behavior of the documents is similar in spirit to Dust and Magnets (DnM) [79]. In DnM, analysts may drag or shake magnets that represent variables in the dataset and watch as relevant documents follow the magnets. However, a major difference between DnM and ugGTM is that when users drag documents (not variables) and watch how the remaining react, they are comparing documents based on all of the variables in the dataset simultaneously. In turn, users may learn which variables are important for comparisons, based on tags within the visualization space.

The interaction is possible because ugGTM gives control to the users of some parameters in the model via the visualization. Let r^* represent the low-dimensional coordinates for a document that an analyst has chosen to drag. Given r^*, we add to the model described in Equations (5.3)–(5.6) by expanding sets g and $\boldsymbol{\Phi}$ so that $g = \{g_1, \ldots, g_K, g^*\}$ and $\boldsymbol{\Phi} = \{\Phi_1, \ldots, \Phi_J, \Phi^*\}$, where $\Phi^* = \exp\{-||r_i - \mu^*||^2/2\sigma^2\}$ and $g^* = \mu^* = r^*$. In turn, we assign the posterior responsibility (Equation (5.6)) that r^* generates d^* via m^* to 1 (where, m^* is defined by Equation (5.4)) so that the mapping between the low- and high-dimensional coordinates for the moving observations is deterministic.

To propagate the effect of moving r^* to the remaining visualization, we take a local regression approach [32] to characterize high-dimensional data $d_i | \{r_i = g^*, m^*\}$ in that we scale $d_i - m^*$ by the square root of function V given scaled distance $\Delta_i = ||d^* - d_i||/c$ so that,

$$\pi\left(d_i | r_i = g^*, W, \Phi\right) = \left(\frac{\beta}{2\pi}\right)^{-p/2} \exp\left\{-\frac{\beta V(\Delta_i)}{2}||d_i - m^*||^2\right\}$$

where c is user-defined; e.g., $V(\Delta_i) = \Delta_i^2$ and $c = 0.5$. In turn, both posterior responsibility estimates (Equation (5.6)) and estimates for $\mathbf{m}$ (Equation (5.4)) change. Let $m_i^{(c)}$ and $m_i^{(u)}$ represent the current and user-adjusted manifold estimates for observation i. We define the BaVA-GTM estimate for the manifold, $m_i^{(c+1)}$, by

$$m_i^{(c+1)} = \delta_i m_i^{(c)} + (1 - \delta_i) m_i^{(u)},$$

where $\delta_i = ||r_i - r^*||/b$ and $b = \max\{||r_1 - r^*||, \ldots, ||r_n - r^*||\}$ so that $\delta_i \in [0, 1]$. This definition for $m_i^{(c+1)}$ controls the visualization so that only the regions of interest respond to user interactions; areas that are distant from the dragged observations do not change.

Parameters $g*, \Phi*, V(\Delta_i), \delta$ and $m^{(c+1)}$ in ugGTM work together in the following way. When a data point d_i is far from d^*, $V(\Delta_i)$ will be large and thus decrease the posterior responsibility (Equation (5.6)) that $r_i = g^*$ generates d_i. Similarly, when d_i is near d^*, the corresponding responsibility will increase. Increases in the responsibility for $r_i = g^*$ will cause the coordinates for r_i to gravitate toward r^*. Thus, analysts may specify constant c in our definition Δ_i, depending upon how many document matches they seek for the moving document. Also, the degree to which the observations gravitate toward r^* is determined by δ and $\mathbf{m}^{(c+1)}$. When the manifold shifts from $\mathbf{m}^{(c)}$ to $\mathbf{m}^{(c+1)}$, the meaning of the visualization space changes, as we demonstrate in our example.

5.2.4 INTERAXIS: STEERING SCATTERPLOT AXES

Another example application that makes use of semantic interaction techniques is a visual analytic prototype called *InterAxis*, shown in Figure 5.11 Kim et al. presented this technique as a means for users to directly manipulate the axes of scatterplots [92]. In traditional scatterplots, the x and y axes of scatterplots are used to show the relationship between two data attributes. The use of two orthogonal axes mapped to data attributes produces a Cartesian space where data objects can be charted. A basic strategy to form these axes in multi-dimensional data visualization is to assign each axis an individual feature or dimension originally given in a data set. For example, plotting temperature over time on the y- and x-axis, respectively, generates a chart that can be used to understand the relationship between these two data attributes. However, this has a severe scalability issue since 2D scatter plots can represent only two data attributes at a time.

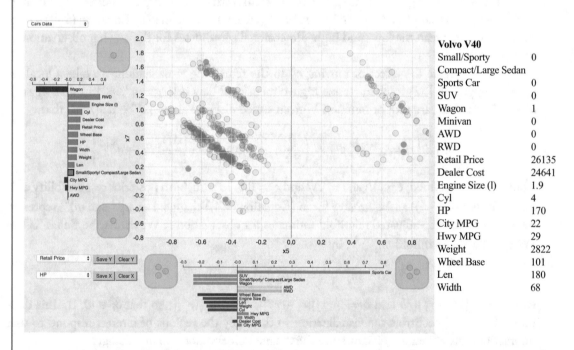

Figure 5.11: Interaxis, developed by Kim et al., lets people use semantic interactions to demonstrate semantically meaningful concepts and map them to axes of scatterplots. For example, people can give examples of cars they like and cars they do not like, place them on opposite sides of an axis, and have the system model the similarity function between the two groups to drive the visualization (adapted from [92]).

Instead, InterAxis provides people the ability to directly manipulate the axes of scatterplots produced via dimension reduction techniques (such as those described in previous sections). People can adjust the axes in two primary ways: directly controlling the attribute contributions, and providing exemplar data points to place on the high or low sides of the axes. The bars next to each axis shows the relative contribution or weight of each data attribute. As a result, the specific axis modified is updated and a new scatterplot is generated based on a linear combination of the attribute weights mapped to the axis.

Alternatively, users can directly interact with the visual representation to make adjustments. For example, if users have the concept of data items they "like" and a few that they "do not like," they can drag those points to the high and low sides of an axis. InterAxis solves for the linear combination of data attributes that best describes this difference, and maps this to the axis (an interaction more similar to the semantic interaction concept compared to directly manipulating the attribute contributions above).

CHAPTER 6

Evaluating Semantic Interaction

Semantic interaction can provide computational support for sensemaking through model steering. This implicit form of model steering can be advantageous for exploratory data analysis, performing model steering on behalf of the user. However, evaluating the effectiveness and accuracy of semantic interaction can be challenging. In addition to visual analytic applications providing users the ability to discover insights about the data, evaluating semantic interaction requires analysis of the evolution of the computational model during analysis. Thus, while existing evaluation methodologies may still apply to the holistic systems developed, additional care should be given to evaluate the integration of semantic interaction.

In this chapter, we present the method and results of a user study analyzing semantic interaction in ForceSPIRE to give an example of how semantic interaction can be directly evaluated.

6.1   METHODOLOGY CONSIDERATIONS

Evaluating visual analytic applications that have semantic interaction integrated requires additional considerations when designing the experiment. In addition to the standard usability and utility metrics that are often used in evaluating visualization applications, researchers also want to directly study the effectiveness of semantic interaction in the context of their applications. In this section, we discuss two high-level concepts that can be used in the evaluation methodology to directly evaluate semantic interaction.

1. Evaluate the **analytic process** of users to understand: 1) the evolution of the computational models being steered by semantic interaction, and 2) the evolution of the visualization co-created between the system and the user.

2. Evaluate the **analytic product** and insights discovered by the users in conjunction with the final state of the visualization and computational model.

6.1.1   EVALUATION OF ANALYTIC PROCESS

The analytic process and discourse that people engage in during visual data exploration is complex. People incrementally build an understanding of the data, augmenting their mental model and understanding of the phenomena they are exploring. As such, one area of evaluation comes through the understanding of how well semantic interaction captured the incremental changes to the mental model of people with the computational model steering in the system.

For example, semantic interaction can be designed to steer a weight vector of data attributes given the user interaction (i.e., to approximate user importance or interest). In this case, each interaction triggers a change to this attribute weight vector. When logged at each increment, this temporal information allows evaluators a better understanding of how well semantic interaction evolved this approximation of user interest over time. The accuracy of this evolution can be studied when used in conjunction with a think-aloud protocol to understand what insights or questions people had at specific times during their investigation.

Similarly, evaluators can study the incremental changes to the visualization. Semantic interaction advocates for using the visualization as a medium through which people perceive insights, and also communicate. Similarly, analytic models can communicate their results, as well as listen and learn from user interactions. Thus, the incremental co-creation of the visualization over time lets evaluators study how well the system and the user are "in sync." Questions that can be answered via such a methodology include: how much effort is spent by the user and by the system to organize the data spatially (when using a spatialization as the visual metaphor)? How well does the user and the system agree on the construction and meaning of the visual metaphor?

6.1.2   EVALUATION OF ANALYTIC PRODUCT

There is often a result that is created from the analytic process. For example, this can be a newly understood and discovered insight about the data, or a visualization that depicts some phenomena used as part of a decision-making process. Similar to the analysis of the process described above, the evaluation here allows for questions regarding the level of agreement between the system and the user in terms of the visualization created. The analytic model is inherently bound to the visual metaphor and encodings. Similarly, users likely discovered specific insights which may be represented in the visualization. One direct application of evaluating semantic interaction is to analyze how well the user's description of the visualization and findings match with the aspects of the data that the steered analytic model emphasizes.

6.2   EXAMPLE: EVALUATING SEMANTIC INTERACTION IN FORCESPIRE

This section presents an example of how the methodological considerations above can be realized in a user study of ForceSPIRE. Below, the study design, data analysis, and results are discussed.

6.2.1   METHOD

This user study investigates the following research questions about the capabilities and benefits of semantic interaction:

1. How well can semantic interaction systematically quantify analytical reasoning based on user interaction as a dynamic entity-weighting scheme?

2. How does the real-time modification of the weighting scheme and adjustment of the spatialization aid users' sensemaking?

3. What was the focus of users while exploring the dataset through semantic interaction? That is, were they focused on adjusting the weighting scheme, or synthesizing information?

4. How does the co-created spatialization map to the users' findings?

We hypothesize that the coupling between the semantic interactions and model updates will create a dynamic weighting scheme that appropriately captures the analytical reasoning of each user throughout his or her investigation. As a result, this weighting scheme will adjust the spatialization, aiding in the co-creation of the layout, where users need not develop the entire layout manually, but also not rely on solely algorithmic generation. During this process, this will help users by adjusting the layout while users read documents and synthesize the information, bringing related documents nearby. Also, we hypothesize that the soft data captured during the analysis will be representative of the analytic product of each user, and therefore the co-created spatialization will be meaningful to the user. Throughout this process, we hypothesize that the users will remain focused on the synthesizing of information, rather than interacting to directly modify the weights of entities.

Equipment

For this study, a large, high-resolution display (shown in Figure 5.1) is used. Such workstations allow users of ForceSPIRE to leverage the additional resolution to show many text documents at full detail, and the additional physical size to provide users with a more embodied analytic experience [4, 20]. This particular workstation is constructed using eight 30-inch displays, driven by a single node, providing a total workspace resolution of 10,240 × 3,200 pixels. The curvature allows easy access to all areas via physical navigation, such as chair rotation [23, 65].

The dataset used for this study is an analysis exercise called Atlantic Storm developed for the purpose of training and evaluating intelligence analysts, as well as analytic tools. The dataset consists of 111 text intelligence reports containing a fictitious terrorist plot. Using LingPipe [1] to extract keywords (i.e., entities) from these documents, 294 unique entities occurring more than once in the dataset were extracted (singletons were removed). The choice to use this dataset is based on the ability to have a realistic dataset, containing a known ground truth against which to compare the findings of the users, while requiring no detailed domain knowledge beyond English reading comprehension and creativity.

Data Collection and Analysis

ForceSPIRE has the ability to log the soft data used for semantic interaction. For the purpose of this study, this gives us a record of every interaction performed by the user, as well as how the system interpreted the interaction in context of the dataset. For example, when a user highlights a phrase, the soft data shows us when the highlight occurred, what the text is, in which document, as well as what entities' weights changed, and what the new weights are.

The users were asked to provide us with verbal feedback throughout their process. In a post-study interview, subjects explained their findings, the resulting spatialization, and insights about their process that may have been missed during the think-aloud protocol. Video recordings and screenshots were also taken during each task for post-study analysis.

Procedure

This study consisted of observing six computer science graduate students. The age of the participants ranged from 27–38, with an average age of 30.

Each participant was given a brief overview of ForceSPIRE, using a practice dataset, for the purpose of making each user familiar with how ForceSPIRE and the supported semantic interactions function. Upon informing us that they were comfortable, each user was given verbal instructions on their task. Each user was given the same initial view of the Atlantic Storm dataset in ForceSPIRE as a starting point. We informed the participants that they had a maximum of 90 minutes to analyze the dataset, after which they would be debriefed regarding their investigation. They were allowed to finish early if they felt they were finished before time expired.

6.2.2   RESULTS

The success of a visual analytic tool hinges on the ability to provide support during the analytic process, as well as a meaningful representation of the user's findings. Thus, the results of this study are presented in terms of the analytic process and product. The analytic process describes how the semantic interactions within ForceSPIRE were used during the analysis, how the corresponding model and spatialization updates benefitted the users, and how the soft data mapped to each user's process. The analytic product details how the findings of each user are represented in the final spatialization, as well as the final weighting of keywords.

Evaluating the Process

Each user's process was different, and thus utilized semantic interaction differently (Table 6.1). However, the analysis of each user's process reveals general usages of each semantic interaction. To address the research questions, we present the analysis of the processes of the users in terms of **usage** (how and when they used each semantic interaction), **reasoning** (what was their purpose for interacting, sensemaking or model steering), **impact on weighting scheme** (how the updated weighting scheme coincided with their reasoning process), and **impact on spatialization** (how did the updating spatialization benefit their analysis).

Semantic interaction usage   Performing a spatial analysis of data focuses around rearranging documents and creating spatial constructs or clusters [4, 27]. As such, **pinning** and **document movement** (both *exploratory* and *expressive*), were the fundamental methods of exploring the

Table 6.1: Semantic interaction counts during each user's analysis

	User						
Interaction	1	2	3	4	5	6	Total
Search	13	32	37	14	38	21	155
Highlight	47	58	12	10	5	0	132
Expressive Movement	45	76	47	62	26	27	283
Exploratory Movement	41	102	64	26	98	43	374
Annotation	3	40	3	0	0	0	46
Total	149	308	163	112	167	91	

dataset. Pinning documents to absolute positions in the spatialization was used to create "spatial landmarks." That is, users pinned a document to a specific location in the layout to create (and maintain) the meaning of a specific region of the spatialization. Based on these landmarks, document movement was used to organize the spatialization based on the user's intuition. For example, User3 pinned a document mentioning "Nassau" in a specific location. From there, he placed other documents related to "Nassau" nearby, and also quickly re-acquired these documents when needed.

Highlighting was used mostly while reading a document to indicate terms or phrases that "stood out." These highlights were beneficial to users to produce visual and cognitive aids. The highlighted phrases (mostly single words and fragments of sentences) helped remind users of what information was important in a document when re-acquiring the document later. User6 was the only user who did not perform any highlighting during his investigation, simply stating that he "did not feel a need to."

Search was used to find other documents containing a term of interest to the user. Generally, users performed a search on keywords for two reasons. First, the unique color assigned to each search provided a quick overview of where in the dataset the term occurred given the current spatial layout. Second, users treated the search window (of which more than one could be opened) as a means to "tag" the space. For example, all of the users commented that leaving multiple search windows open and pinned to specific locations was an effective way to recall the meaning of that specific position in the spatialization.

With the exception of User2, **annotation** was rarely used. User2 said that he enjoyed the ability to "add personalized notes to important documents." In his case, ForceSPIRE detected 23 entities in his annotations (that were not in the dataset), including entities such as "irrelevant," "suspicious," and "revisit" (extracted from a note stating "should revisit this later"). During his investigation, he also found it useful to track what documents he found important by scanning the workspace and seeing which documents had the yellow notes window visible. Thus, annotations can be helpful to some users, while others prefer to utilize other interactions to support their analysis.

Aiding the sensemaking process   The primary benefit for sensemaking provided by ForceSPIRE was aiding the user in adjusting the layout by bringing related information nearby. Each semantic interaction in ForceSPIRE is tightly coupled with the dynamic weighting scheme used by the force-directed model responsible for generating the spatial layout. As such, ForceSPIRE responds to each interaction via updating the spatialization as a result of the updated weighting scheme.

Each user's process involved multiple stages of the investigation, including exploring specific leads (e.g., a person, place, etc.) and hypotheses regarding the dataset. As such, it is important for semantic interaction to allow a flexible entity weighting scheme to support exploring each of these aspects during different times of the investigation. For example, while a user investigates information regarding the entity "Atlanta," the weight of entities similar to (and including) "Atlanta" should increase. If the user chooses to investigate "weapons" at a later time, the weighting scheme should reflect this change. The challenge, then, comes in supporting the rapid and fluid change of what is currently being investigated by a user through rapidly changing keyword weights, while maintaining a history of the previously emphasized keywords.

For example, Figure 6.1 shows the temporal history of User4's keyword weighting during his analysis. The patterns and trends observed in User4's analysis of the soft data is also representative of the other users' history. One can see that approximately two minutes into the investigation, the entity "package" was created. The creation occurred while User4 read a document and found the phrase "carefully wrapped package" important, and thus highlighted it. The effect on the layout was that documents containing the entity "package" were brought closer. He did not immediately switch to reading those documents, instead continued to read the document while other related documents came nearby. This strategy was found in other users' processes also. "It was nice to see what documents would come near while I was reading and highlighting," User1 told us after his investigation. He continued to tell us that he would notice other documents coming closer, but would "continue reading and highlighting until I finished that document, then decide where to go next depending on what's close by." Upon finishing reading the document, User4 pinned it, and chose the closest document to continue his investigation. This document was one related to "package," and important to the plot. He continued reading three more documents containing "package," highlighting other phrases that contained the term. As such, the term continued to increase in weight, and related documents continued to form more tightly around the one that was pinned.

Figure 6.1 also shows instances when User4 explored other potentially relevant information. For example, at 36 minutes into his analysis, he informed us that he wanted find out more information regarding events in "Chicago." This stemmed from reading a document that mentioned "Chicago." He highlighted the single word, and immediately pinned the document in a specific location, away from other documents. Then, he searched for the term "Chicago," and

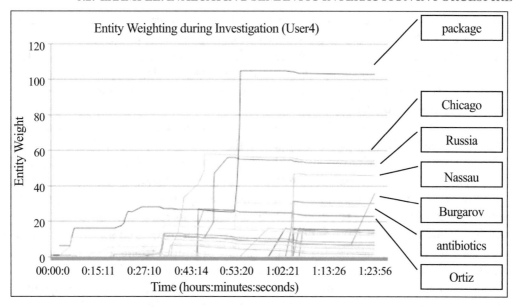

Figure 6.1: User4's entity weighting over the duration of his analytic process. Semantic interactions in ForceSPIRE adjusted the weights of entities to coincide with his investigation of multiple hypotheses. As a result, the layout adjusted incrementally with each interaction.

placed the search window next to the pinned document. Then, he opened and read some of the documents containing Chicago (they were highlighted teal from the search) that came closer. The first two documents he read he placed near the first pinned document. As a result, the weighting of "Chicago" increased, but so did the weights of other related entities, such as "Russia," "weapons," and "Panama City." This occurred because the documents he dragged near the pinned document containing "Chicago" were not similar based on only "Chicago," but those other entities as well. As a result, more documents came closer that did not contain strictly "Chicago," but were related. In this case, he read some of the documents containing "Russia" and "weapons" and moved them into another, separate location. Again, ForceSPIRE responded by moving the documents more similar to "Russia" and "weapons" into that location, rather than near the location regarding "Chicago." This benefitted the user as he noticed how some documents remained in the middle of the two areas, showing relevance to both topics. These were documents connecting these two, and important to the plot.

Toward the end of his investigation (approximately one hour into it), he focused on tying all the pieces of evidence he had collected together. He did so through exploring the relevance of "Bahamas" and "Nassau." He did so primarily through small, local movements within an area specific to each. He arranged the documents within the region to reflect a sequence of events related to transportation of a package. In addition to the weight increase of those entities, he discovered the relevance of some of the key persons involved in the suspicious activity (i.e., "Odeh," "Hanif," and others). He found an important document detailing how some of the weapons (which he investigated earlier) were possibly being transported by students funded through a suspicious scholarship fund. The history of his weighting scheme reflects each of these hypotheses and branches during his investigation, as indicated in Figure 6.1.

In addition to increasing the weighting of entities that were relevant to the investigation, semantic interaction also reduced the weight of entities that were not. ForceSPIRE does so through "decaying" the weights of previously emphasized keywords over time. That is, as other keywords are emphasized via various semantic interactions, weights of previously emphasized keywords will begin to decay. Therefore, if they are never investigated again, they will eventually return to their lower weight, but if they are revisited again later, they will increase in weight again. At times, this resulted in the weights of those entities going to zero (thus having no impact on the spatial layout). Across all users' processes, the average number of entities where this occurred at least once is 245 (out of the 294 unique entities initially extracted by ForceSPIRE). While these entities did not have an impact on the spatial layout when their weight was set to 0, subsequent semantic interactions continued to use these entities to measure similarity. As a result, entities that may not have been relevant during the early stages of an investigation and were relevant to a later hypothesis being explored, saw their weight increased. An example from User4 is the term "Nassau," which was not relevant to his investigation until approximately 54 minutes into the study, where the weight increased from zero (as shown in Figure 6.1). This happened as a result of him dragging one document close to another on the basis of both being about an event in the "Bahamas." ForceSPIRE interpreted this similarity, but also found these documents similar because of "Nassau," increasing the weight of the term and bringing those related documents nearby.

Semantic interaction aided two users from this study in **creating a "junk pile"** (i.e., a collection of documents that are not relevant to the main plot, and are thus placed in a location away from the relevant information). As these two users placed more information into the same cluster that they referred to as "junk," ForceSPIRE calculated the similarity between the documents being placed into this cluster and increased the weight of those entities. "Look! It's moving other junk into my junk pile for me," User1 remarked. However, he was skeptical of the system's ability to detect irrelevant documents, so he opened and read a few of them as they moved closer. Some he agreed with being junk and left them in the junk cluster, while others he moved near other pinned documents in the spatialization. By doing so, he continued to improve ForceSPIRE's ability to detect irrelevant documents. When asked about this experience after the study, he told us that the more he interacted with the layout (including his "junk pile"), the more pleased he became with the metrics for determining junk, and the "more [he] trusted it."

An important capability in ForceSPIRE is the **steering of the entity extraction algorithm** for generating additional entities during an investigation. Entities can be added to the system through semantic interaction, which was critical to the ability to capture and infer the users'

Table 6.2: Number of entities added via semantic interaction during each user's investigation. The majority of these new entities (92%) maintained a weight above 0 throughout their process.

	User					
	1	2	3	4	5	6
Entities Added	43	62	35	13	15	10
Weight > 0	38	54	35	13	14	10

reasoning processes. While the entity extraction algorithm in ForceSPIRE managed to extract 294 unique entities, each user found additional entities that were relevant to their analysis. Table 6.2 shows the number of entities created as a direct outcome of semantic interaction. Of these, most (92%) maintained a weight greater than zero throughout the investigation. This shows that not only was it important for users to steer the weighting of existing, extracted entities, but also to steer the entity extraction algorithms to generate additional entities. For example, User3 highlighted the phrase "he has students now in the USA," which was passed through a more aggressive entity extraction algorithm, and detected "students" as an entity. This entity was important to the user's findings, as well as highly weighted in the model (Table 6.3).

Pinning and un-pinning documents was used not only to place meaningful documents in absolute positions in the workspace, but also to check if the current weighting model would place the document in another region (or into another cluster). For example, three of the users commented that they un-pinned a document to see where it went after it had been pinned for a long time. They were interested in other possible topics it might relate to. If nothing particularly interesting was found, they returned the document to the previous location and pinned it again. However, often users found relationships between these documents and other clusters, and typically either left the document un-pinned, or pinned it in a different location from where it was pinned previously. For instance, User1 found that he had a document pinned from very early in his investigation that referred to the Freeport Star Hotel. When he un-pinned it, he saw it go near other documents about the Bahamas and Nassau, which helped him make the connection about the events happening in that area.

In general, users emphasized the importance of observing the spatialization adjust incrementally. That is, to notice the change in relative distances between documents as a result of the highlighting they did while reading, searching, etc. Such exploration can be found in other tools, such as VIBE [52] or Dust&Magnet [79], where users can place "points of interest" corresponding to keywords in specific locations, and observe how the spatial layout adjusts given those keywords and locations.

Users did not treat the semantic interactions as a means to directly manipulate entities. That is, they interacted as a means to synthesize the information. For instance, based on their comments, highlighting was performed not to pass a phrase through a more aggressive entity

Table 6.3: Each user's top five entities, collected both from the user's debriefing (user), and based on the final entity weighting (model). Underlined entities indicate a match between the user and model. Bold entities were entities added to the model as a result of semantic interaction during the analytic process (i.e., missed by the initial entity extraction).

1 (user)	1 (model)	2 (user)	2 (model)	3 (user)	3 (model)	4 (user)	4 (model)	5 (user)	5 (model)	6 (user)	6 (model)
diamonds	diamonds	package	Nassau	explosives	Nassau	diamonds	**package**	Al Queda	Nassau	diamonds	**weapons**
scholarship	**weapons**	Hanif	Hanif	weapons	**students**	Nassau	Chicago	Caribbean	Miami	antibiotics	**diamonds**
jihad	**graduate**	antibiotics	Freeport	Nassau	**weapons**	Burgarov	Russia	Russia	Freeport	Ortiz	Nassau
weapons	Jamal	diamonds	Miami	students	**scholarship**	antibiotics	Nassau	Hanif	Apple St.	Bahamas	**graduate**
Freeport	Nassau	Nassau	**package**	scholarship	Jamal	package	Burgarov	Odeh	**weapons**	Hijazi	Hijazi

extraction algorithm, but to emphasize a part of the text as being important, so as to be able to find it again more easily later. "Oh, that's important ... [I] might need to come back to [it]," one user stated while highlighting a phrase. None of the users found the need to directly manipulate entities (e.g., adjust the weights, add, remove) via the "Entity Viewer" (shown in Figure 6.3). All users were shown this feature in the training, but none found it necessary to use during their actual investigations. These results evidence that semantic interaction properly coupled the semantic interactions with model updates, to the extent that users never felt the need to do so directly. This contrasts with the intended usage of other tools, such as IN-SPIRE [53], where model steering occurs via direct parameter manipulation on the part of the user. With ForceSPIRE, users were successfully able to focus on the synthesis necessary for sensemaking, while the parameter adjustments occurred systematically in accordance with their analytic reasoning.

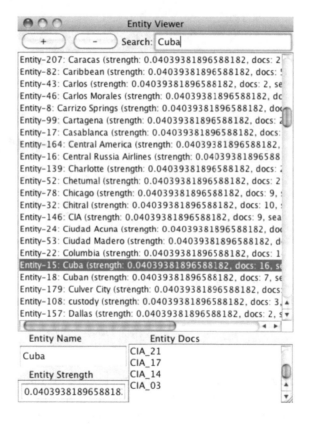

Figure 6.2: The "Entity Viewer" in ForceSPIRE allows users direct control over the weights of entities, adding entities, and removing them. With semantic interaction, this view was never needed.

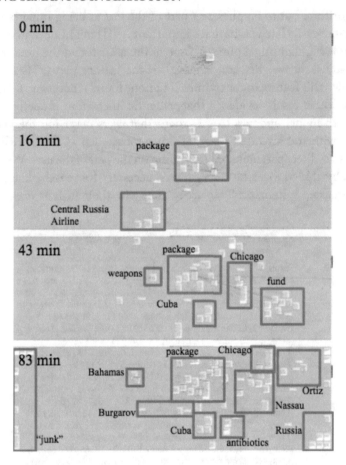

Figure 6.3: The progression of the spatialization over time for User04. The annotations (labels and red region boundaries) were added after the study. They represent the meaning of the regions of the space, as indicated by the user during his process.

Evaluating the Product

At the conclusion of each trial, we asked the user to describe his findings (both in terms of what information is relevant to the suspicious plot, which all users found, and information that was not). ForceSPIRE remained visible during this debriefing, but we asked users to not interact with the tool, but simply use the final layout as a means to help describe their findings. The analytic product with regard to this study refers to the final spatial layout in ForceSPIRE, as well as the user's debriefing after the study. We analyze this information in addition to the final weighting scheme (i.e., soft data) at the conclusion of the analysis. Compared to the known ground truth of the dataset, each user found the suspicious activities, with varying amounts of detail to support the findings. Thus, the analysis of the product here is with regard to ForceSPIRE's ability to create synergy between what the system and the user knows (rather than compared to the ground truth).

The final weighting schemes at the conclusion of each user's investigation served as a good approximation of the keywords relevant to their findings (see Table 6.3). This table shows the top five keywords from each user's debriefing and the top five entities based on entity weight, labeled "user" and "model" respectively. The entities from the debriefing were given to us by the user as part of their debriefing to represent their findings. The entity weights were obtained by taking the top five highest weighted entities in the final soft data state. The entities highlighted in bold are entities that were not initially extracted, but added to the model through the user's semantic interaction during the study.

These results reveal that 47% of the entities match directly from the user's findings to the highly weighted entities. In addition, eleven of the highly weighted entities were added to the model as a result of semantic interaction. Of these eleven, seven were important based on the debriefing of the user (i.e., they matched with the top five entities given to us by the user). Therefore, not only were these entities added as a result of semantic interaction, but they were also relevant to the user's findings.

In some cases, there is not a direct match between the entities obtained during the debriefing and the entity weighting. For example, User5's entities show no direct matches. However, a more sophisticated entity correlation algorithm may find connections between entities such as "Caribbean" and "Nassau," "Freeport," "Miami," and "Apple St." (an address in Nassau in this dataset). Thus, even when there are not direct matches, we find that the higher weighted entities provide a good estimate of characteristics of the dataset users found important and relevant to their investigation. In fact, this indicates that the system was successfully able to interpret the user's reasoning within the context of the actual data. That is, the system identified keywords relevant to the user's process that the user did not think of or at least had used other words in place of. This suggests that the system fulfills the needs of incremental formalism and ill-defined user-generated clustering [27].

Spatialization co-creation   One of the goals of semantic interaction is to properly steer the underlying model to allow for co-creation of the spatialization between the user and the model. As such, we hypothesized that some documents (and search windows) would be pinned by the user to maintain their absolute position in the workspace, and others would be un-pinned so that the model could determine their position based on the current entity weighting. Table 6.4 shows the number of pinned and un-pinned documents and search windows in the final layout produced by each user.

In the final layouts, 32% of the documents were pinned. Based on the debriefing, the users informed us that some documents were pinned to maintain their absolute position in the workspace for the purpose placing meaning into the workspace. However, others were at times pinned for more detailed adjustments to cluster layout. For example, User2 commented that he would pin documents in a cluster so that he could create detailed spatial structures within a cluster

Table 6.4: Pinned and unpinned documents and search windows in each user's final spatialization

		User					
		1	2	3	4	5	6
Docs.	Pinned	34	41	44	28	32	34
	Un-Pinned	77	70	67	83	79	77
	Detail	85	88	89	36	40	49
	Minimized	26	23	22	75	71	
Search	Pinned	9	19	31	2	16	5
Window	Un-Pinned	0	0	0	1	1	0

(e.g., a timeline). ForceSPIRE currently does not well support such multi-scale spatial layouts, but feedback from users suggests doing so in the future.

The majority of search windows were pinned (98%). Users treated them as "tags" for their spatialization, as searches were performed on entities. The only two users who had one search window un-pinned (User4 and User5) explained that they preferred to have it "float" to get an idea of where the documents are that related to that term.

For example, User4's progression of the co-created spatialization can be seen in Figure 6.3. After 16 minutes, his layout was made up of two main clusters: one about "package" and one about the "Central Russian Airline." Then, as his investigation continued, he learned more about the dataset (e.g., a suspicious "fund," documents about "weapons," etc.). At the completion of his trial (83 minutes), he was aware of much more detail regarding the dataset, such as events happening in "Nassau" regarding the "package," a suspicious person named "Ortiz," and a unrelated plot in "Russia." Additionally, the layout placed a collection of documents along the far left, which the user told us was "junk." These results indicate that the spatialization was successfully co-created (based on the coinciding weights, shown in Figure 6.3), and maintained meaning for the user. As such, not only did the weights reflect the analytic reasoning of the user, but the spatialization seemed to successfully reflect the shared knowledge between the user and the system at each stage.

CHAPTER 7

Discussion and Open Challenges

Semantic interaction has shown contributions in specific areas of visual analytics, as described in the previous sections. Applications have been developed that provide evidence that this form of user interaction provides utility through new forms of people interacting with analytic models. Further, user studies have shown the benefit of semantic interaction, in that people can inject domain expertise into the analytic models to help drive the computation. With these advances, there are also additional open questions and challenges going forward for visual analytics with regard to advancing user interaction. Below some of those are discussed.

7.1   USER INTERACTION FOR VISUAL ANALYTICS

While the existing research on semantic interaction reveals important opportunities and findings, there is much more work that can be done to further frame and solidify the understanding of user interaction for visual analytics. This need for understanding comes from an important shift in how we think about user interaction in these systems. Interaction is no longer just an ephemeral action that moves visualizations from one state to the next. Instead, the sequences of interactions are systematically interpreted and analyzed to evolve and adapt the systems during use. This opens up many research questions, including:

- What input and feedback parameters for interactions are important to users?

- What tradeoffs exist between designing algorithms to minimize low-dimensional distances between information, vs. maintaining absolute locations of information in the spatialization?

- How can interactions be designed to allow both entity weighting and document weighting, and what impact should those two weighting mechanisms have on the layout?

- How can we leverage large display, multi-touch interactions to enable users to implicitly or explicitly specify a wider range of parameters to inform the system of their analytical reasoning?

- How can we design effective means of providing visual feedback to users about what the system has learned from the user's interactions?

In a spatialization, the spatial positioning of information is one of the primary visual encodings for the information. Distances between documents can imply a similarity measure, while absolute locations of information can serve as a "landmark" for themes and concepts within the spatialization. These two encodings can inform the system of different, yet equally important information. For instance, generating clusters of documents can signal a similarity between the collection of documents, from which the statistical model can mine the important information creating the similarity. Instead, the absolute location of a document in the spatialization can inform the system that the user has linked a particular theme with the specific, persistent location. Therefore, the goal is to provide algorithms that not only aim to optimize the low-dimensional distances between pieces of information, but also attempt to maintain the persistence of the absolute locations provided by the user.

The input parameters for users will primarily control the "weighting" (or importance/relevance) of either entities or documents. Users have the ability to control the weights of entities through interactions such as searching, highlighting, etc. Changing the weights on entities will allow the system to update the spatial layout, emphasizing those entities. Weighting documents has a different effect. When users emphasize an important document, the weight of that document increases. As a result, that document becomes more persistent, and the location of the document is more resistant to change from the layout algorithm.

Similarity as a measure includes many parameters to help inform the system of the details of the similarity between pieces of information. For example, users can specify which information is being compared, their level of confidence, the specific entities causing the similarity, and more. However, forcing the user to specify each of these can be overbearing, especially in the early stages of the analysis. What we know from *incremental formalism* [63] is that users gain insight incrementally over the course of their investigation. That is, information to populate each of those parameters may not be known initially, instead only realized later in the analysis. It is important for the system to allow this incremental expressiveness in terms of parameter specification. This allows the analytic process to proceed, and enables users to provide the system with only the information they posses at their current stage.

One promising interaction modality for enabling users to specify these parameters is through the use of large display, multi-touch surface. Instead of a traditional mouse and keyboard interface, where parameters need to be specified by modifier keys, menus, pop-up dialog boxes, etc., multi-touch allows more modifiers through allowing more simultaneous points of input. We plan to explore how gestures can be designed to accommodate the specification of these parameters [51]. For example, while a user moves a document with his right hand, he can use his left hand to specify which documents he intends to compare the document to. Further, specific gestures can be used to specify parameters such as the level of granularity of the comparison (e.g., entity level, document level, etc.).

Another important aspect of this research task will be to explore effective means of providing visual feedback to users. In a spatialization, the challenge is to determine the correct balance between visual and mathematical feedback, as well as correctly determining the appropriate level of feedback. Visual feedback can be presented by updating the locations of information, changing the color of links between pieces of information information, etc., to signify the system's response to the user's domain knowledge. We plan to experiment with different detail levels of the feedback, such as presenting feedback on the weighting of documents, down to the level of showing the weights of entities. It is also important to allow users to react to the feedback in case the system misinterpreted the analytical reasoning of the users. For example, feedback on the entities causing the similarity within a cluster can be represented using a tag cloud, where entities with strong weights are represented larger. Users can correct the similarity in this visual metaphor by changing the size of the entities, or removing them completely. The form of the visual feedback should correspond to the user's understanding of the space. For example, document similarities could be expressed through proximity, while entity importance could be expressed through highlighting. One outcome of this approach is that important information should become more visually salient to the user.

7.2   EFFECTS OF SEMANTIC INTERACTION ON ANALYTIC PROCESS

One approach is to observe users analyzing a textual dataset using a prototype, ForceSPIRE, which utilizes semantic interaction (and thus captures soft data throughout the analysis). To provide a realistic intelligence analysis task with a known ground truth to evaluate the performance of the users against, we plan to use one of the VAST Challenge Datasets [58]. One can collect both qualitative results in terms of the behavior and processes of the users, as well as quantitative results such as how close their findings match the known ground truth, and each of the user's soft data. As soft data gives us a quantified form of the user's domain knowledge and intuition regarding the dataset, we can analyze that information to establish a level of bias. If the goal of a successful visual analytic system is to balance human intuition with statistical models, the balance between hard data and soft data can provide one such measure. The questions to answer are what is the balance, and how much intuition is needed (and how much is too much, or too little)? The soft data will give an idea of how much the user has guided the model through weighting of entities and documents, and how much the user has constrained the model through pinning documents in specific locations.

7.3   DIFFERENTIATING BIAS FROM INTUITION

The concept of human-in-the-loop model steering is often associated with the assumption that people will "bias" the computation in some way. It is assumed that the initial result generated by the analytic model, given the data, is somehow "optimal" by quantitative measures used to evaluate computation (e.g., stress, entropy, etc.). Instead, semantic interaction (as well as human-in-the-loop concepts) advocate for people learning about the model, and contributing domain expertise and subjective preferences to the model. Through exploration of different model permutations, people ideally learn about the model and the data, gaining insight into both. Because semantic interaction also contributes soft data to the models, they actually become more rich and perhaps more accurately suited to the domain and user.

However, the concept of "bias" is still very relevant when considering model steering and user interaction. Prior studies of people analyzing data have shown that people can exhibit many different forms of bias, often without realizing they are [35]. For example, confirmation bias or anchoring can lead to people latching onto the first finding or insight found, and not considering enough alternatives during the analysis. Along with the other forms of bias, these can ultimately be detrimental to data analysis.

With semantic interaction, systems have the potential to observe specific forms of cognitive bias. Since there is a quantitative model of how the user's input changes the analytic model at each temporal step of the analysis, the analytic discourse can be quantitatively analyzed. For example, confirmation bias or anchoring could be determined via heuristics that observe data attributes that were initially highly weighted by the user, and never reconsidered later. When made evident to the user, this could result in the analyst either realizing his or her own bias and combating it, or the re-affirmation that the specific data attribute should be highly weighted in the context of the user's task, domain, etc. This opens several additional research directions around how to perform mixed-initiative tasks in the context of bias and model steering for visual analytics.

7.4   ADDITIONAL VISUAL REPRESENTATIONS AND INTERACTIONS

Semantic interaction in this work was explored using primarily textual datasets visualized using spatializations. However, semantic interaction can generalize to other visual representations, and steer the underlying mathematics of a variety of visualizations. The fundamental principles can be applied to these visualizations by designing interactions that are not designed based on directly manipulating parameters, but instead adhere to the visual sensemaking process and reasoning of users given the visual representation. Further, these different visual representations may have unique styles of interactions that they afford, and thus other types of semantic interactions may exist for them which are not afforded in a spatial workspace.

CHAPTER 8

Conclusion

The field of visual analytics is at an exciting, yet critical moment. Data sizes and complexities are increasing; machine learning and data mining techniques are being created and improved upon at an impressive rate. Increasingly people look to data as a means to describe or understand specific phenomena. Visual analytics can provide people with such capabilities, creating insights into new domains and areas where data is becoming available. However, to continue to enable such insights, a closer relationship between the computational capabilities and human reasoning processes must be explored. Through advancing user interaction techniques for visual analytics, computation and cognition can be more closely coupled.

Semantic interaction presents one approach to creating this coupling between data analytics and human reasoning processes. Systems that utilize semantic interaction learn about the interests and domain expertise of people to create more meaningful analytic models, and in turn more insightful visualizations. Semantic interaction allows people to explore their data using interactions native to the visual metaphor and data representation—instead of interactions tailored toward the analytic model parameters. Thus, visual analytics can leverage the expertise and reasoning of domain experts, rather than requiring these users to learn about data analytic models in order to effectively use the system.

We are in the early stages of understanding the entirety of the semantic interaction concept space. While we have shown applicability in steering specific models, and leveraging interactions native to spatializations and textual data, there are likely additional possibilities and combinations. Exploring these possibilities can lead to a more well-established "science of interaction," and ultimately allow for the creation of more powerful and usable data analysis tools. Such tools can enable people across society to become more well-equipped in the current data-driven era.

References

[1] *Alias-i. 2008. LingPipe 4.0.1*, 2008. 61

[2] Alonso, O., Gertz, M., and Baeza-Yates, R. Clustering and exploring search results using timeline constructions. *Proc. of the 18th ACM Conference on Information and Knowledge Management*, Hong Kong, China, 2009. ACM, pages 97–106. DOI: 10.1145/1645953.1645968. 19

[3] Alsakran, J., Chen, Y., Zhao, Y., Yang, J., and Luo, D. STREAMIT: Dynamic visualization and interactive exploration of text streams. *IEEE Pacific Visualization Symposium*, 2011. DOI: 10.1109/pacificvis.2011.5742382. 20

[4] Andrews, C. *Space to Think: Sensemaking and Large, High-Resolution Displays*. Virginia Tech, Blacksburg, 2011. 61, 62

[5] Andrews, C., Endert, A., and North, C. Space to think: Large, high-resolution displays for sensemaking. *CHI*, 2010, pages 55–64. DOI: 10.1145/1753326.1753336. 18, 19, 24, 26

[6] Andrews, C., Endert, A., and North, C. VAST 2010 challenge: Analyst's workspace. *IEEE VAST Extended Abstracts (Contest Submission)*, 2010.

[7] Andrews, C., Endert, A., Yost, B., and North, C. Information visualization on large, high-resolution displays: Issues, challenges, and opportunities. *Information Visualization*, 10(4), 2011, pages 341–355. DOI: 10.1177/1473871611415997.

[8] Ball, R., North, C., and Bowman, D. Move to improve: Promoting physical navigation to increase user performance with large displays. *CHI 2007*, San Jose, California, 2007. ACM. DOI: 10.1145/1240624.1240656.

[9] Blake, C. and Merz, C. J. *{UCI} Repository of Machine Learning Databases*, 1998. 49

[10] Buja, A., Swayne, D. F., Littman, M., Dean, N., Hofmann, H., and Chen, L. Interactive data visualization with multidimensional scaling. *Journal of Computational and Graphical Statistics*, 17(2), 2008, pages 444–472. DOI: 10.1198/106186008x318440. 51

[11] Callahan, S. P., Freire, J., Santos, E., Scheidegger, C. E., Silva, T., and Vo, H. T. VisTrails: visualization meets data management. *Proc. of the 2006 ACM SIGMOD International Conference on Management of Data*, Chicago, IL, 2006. ACM. DOI: 10.1145/1142473.1142574. 25

[12] Card, S. K., Mackinlay, J. D., and Shneiderman, B. *Readings in Information Visualization: Using Vision to Think*. Morgan Kaufmann Publishers Inc., 1999. 12

[13] Chang, R., Ghoniem, M., Kosara, R., Ribarsky, W., Yang, J., Suma, E., Ziemkiewicz, C., Kern, D., and Sudjianto, A. WireVis: Visualization of categorical, time-varying data from financial transactions. *Proc. of the 2007 IEEE Symposium on Visual Analytics Science and Technology*, 2007. IEEE Computer Society, pages 155–162. DOI: 10.1109/vast.2007.4389009. 12

[14] Christopher, M. B. *GTM: The Generative Topographic Mapping*, 1998. DOI: 10.1162/089976698300017953. 51

[15] Cowley, P., Haack, J., Littlefield, R., and Hampson, E. Glass box: Capturing, archiving, and retrieving workstation activities. *Proc. of the 3rd ACM Workshop on Continuous Archival and Retrieval of Personal Experiences*, Santa Barbara, California, 2006. ACM, pages 13–18. DOI: 10.1145/1178657.1178662. 25

[16] Dempster, A. P., Laird, N. M., and Rubin, D. B. *Maximum Likelihood from Incomplete Data Via the EM Algorithm*, 1977. 52

[17] Dou, W., Jeong, D. H., Stukes, F., Ribarsky, W., Lipford, H. R., and Chang, R. Recovering reasoning processes from user interactions. *IEEE Computer Graphics and Applications*, 29, 2009, pages 52–61. DOI: 10.1109/mcg.2009.49. 12, 28

[18] Drucker, S. M., Fisher, D., and Basu, S. Helping users sort faster with adaptive machine learning recommendations. *Proc. of the 13th IFIP TC 13 International Conference on Human-computer Interaction—Volume Part III*, Lisbon, Portugal, 2011. Springer-Verlag, pages 187–203. DOI: 10.1007/978-3-642-23765-2_13. 32

[19] Elmqvist, N., Moere, A. V., Jetter, H.-C., Cernea, D., Reiterer, H., and Jankun-Kelly, T. Fluid interaction for information visualization. *Information Visualization*, 10(4), 2011, pages 327–340. DOI: 10.1177/1473871611413180. 12

[20] Endert, A., Andrews, C., Bradel, L., Zeitz, J., and North, C. *Designing Large High-Resolution Display Workspaces*, 2012. DOI: 10.1145/2254556.2254570. 61

[21] Endert, A., Andrews, C., Fink, G. A., and North, C. Professional analysts using a large, high-resolution display. *IEEE VAST Extended Abstract*, 2009. DOI: 10.1109/vast.2009.5332485.

[22] Endert, A., Andrews, C., and North, C. Visual encodings that support physical navigation on large displays. *Graphics Interface*, Virginia Tech, 2011.

[23] Endert, A., Fiaux, P., Chung, H., Stewart, M., Andrews, C., and North, C. Chair-Mouse: Leveraging natural chair rotation for cursor navigation on large, high-resolution displays. *Proc. of the 2011 Annual Conference Extended Abstracts on Human Factors in Computing Systems*, Vancouver, BC, Canada, 2011. ACM, pages 571–580. DOI: 10.1145/1979742.1979628. 61

[24] Endert, A., Fiaux, P., and North, C. Semantic interaction for sensemaking: Inferring analytical reasoning for model steering. *IEEE Conference on Visual Analytics Science and Technology*, 2012. DOI: 10.1109/tvcg.2012.260. 29

[25] Endert, A., Fiaux, P., and North, C. Semantic interaction for visual text analytics. *CHI*, 2012. DOI: 10.1145/2207676.2207741. 3, 25, 33

[26] Endert, A., Fiaux, P., and North, C. Unifying the sensemaking loop with semantic interaction. *IEEE Workshop on Interactive Visual Text Analytics for Decision Making at VisWeek 2011*, Providence, RI, 2011.

[27] Endert, A., Fox, S., Maiti, D., Leman, S. C., and North, C. The semantics of clustering: Analysis of user-generated spatializations of text documents. *AVI*, 2012. DOI: 10.1145/2254556.2254660. 19, 62, 71

[28] Endert, A., Han, C., Maiti, D., House, L., Leman, S. C., and North, C. Observation-level interaction with statistical models for visual analytics. *IEEE VAST*, 2011, pages 121–130. DOI: 10.1109/vast.2011.6102449. 29, 30, 31, 41, 45

[29] Fruchterman, T. M. J. and Reingold, E. M. Graph drawing by force-directed placement. *Software: Practice and Experience*, 21(11), 1991, pages 1129–1164. DOI: 10.1002/spe.4380211102. 35, 40

[30] Gotz, D. *Interactive Visual Synthesis of Analytic Knowledge*, 2006. DOI: 10.1109/vast.2006.261430. 25

[31] Guber, D. Getting what you pay for: The debate over equity in public school expenditures. *Journal of Statistics Education*, 7(2), 1999. 47

[32] Hastie, T., Tibshirani, R., and Friedman, J. H. *The Elements of Statistical Learning*. Springer, 2003. DOI: 10.1007/978-0-387-21606-5. 55

[33] Heer, J. *Prefuse Manual*, 2006. 10, 25

[34] Heer, J., Mackinlay, J., Stolte, C., and Agrawala, M. Graphical histories for visualization: Supporting analysis, communication, and evaluation. *IEEE Transactions on Visualization and Computer Graphics*, 14(6), 2008, pages 1189–1196. DOI: 10.1109/tvcg.2008.137. 25

[35] Heuer, R. *Psychology of Intelligence Analysis*, 1999. 76

[36] Hossain, M. S., Andrews, C., Ramakrishnan, N., and North, C. Helping intelligence analysis make connections. *Workshop on Scalable Integration of Analytics and Visualization*, San Fransisco, 2011.

[37] Hossain, M. S., Gresock, J., Edmonds, Y., Helm, R., Potts, M., and Ramakrishnan, N. Connecting the dots between PubMed abstracts. *PLOS One*, 7(1), 2012, e29509. DOI: 10.1371/journal.pone.0029509.

[38] House, L., Leman, S. C., and Han, C. Bayesian Visual Analytics (BaVA). *In Revision, Technical Report: FODAVA-10-02*, 2010. http://fodava.gatech.edu/node/34. DOI: 10.1002/sam.11253. 46

[39] Jeong, D. H., Ziemkiewicz, C., Fisher, B., Ribarsky, W., and Chang, R. iPCA: An interactive system for PCA-based visual analytics. *Computer Graphics Forum*, 28, 2009, pages 767–774. DOI: 10.1111/j.1467-8659.2009.01475.x. 13, 15, 20

[40] Jolliffe, I. *Principal Component Analysis*. John Wiley & Sons, Ltd, 2002. DOI: 10.1002/9781118445112.stat06472. 45

[41] Kaban, A. A scalable generative topographic mapping for sparse data sequences. *International Conference on Information Technology: Coding and Computing (ITCC'05)*, 2005. DOI: 10.1109/itcc.2005.34. 19

[42] Karen, A. Statistical interpretation of term specificity and its application in retrieval. *Journal of Documentation*, 28, 1972, pages 11–21. DOI: 10.1108/eb026526. 36

[43] Kohonen, T., Kaski, S., Lagus, K., Salojarvi, J., Honkela, J., Paatero, V., and Saarela, A. Self organization of a massive document collection. *Transactions on Neural Networks*, 11(3), 2000. DOI: 10.1109/72.846729. 51

[44] Leman, S. C., House, L., Maiti, D., Endert, A., and North, C. *A Bi-directional Visualization Pipeline that Enables Visual to Parametric Interation (V2PI)*. NSF FODAVA Technical Report (FODAVA-10-41), 2011.

[45] Liu, J., Brown, E. T., and Chang, R. Find distance function, hide model inference. *Poster at IEEE Conference on Visual Analytics Science and Technology*, 2011. DOI: 10.1109/vast.2011.6102478. 29, 30

[46] MacQueen, J. Some methods for classification and analysis of multivariate observations. *Proc. of the Berkeley Symposium on Mathematical Statistics and Probability*, 1, 1967, pages 281–297. 53

[47] Marshall, C. C., Frank, M., Shipman, I., and Coombs, J. H. VIKI: spatial hypertext supporting emergent structure. *Proc. of the 1994 ACM European Conference*

on Hypermedia Technology Edinburgh, Scotland, 1994. ACM, pages 13–23. DOI: 10.1145/192757.192759. 18

[48] Marshall, C. C. and Rogers, R. A. Two years before the mist: experiences with Aquanet. *Proc. of the ACM Conference on Hypertext*, Milan, Italy, 1992. ACM, pages 53–62. DOI: 10.1145/168466.168490. 18

[49] McLachlan, G. J. and Basford, K. E. *Mixture models. Inference and Applications to Clustering*, Dekker, 1988.

[50] North, C., Chang, R., Endert, A., Dou, W., May, R., Pike, B., and Fink, G. Analytic provenance: process+interaction+insight. *Proc. of the 2011 Annual Conference Extended Abstracts on Human Factors in Computing Systems*, Vancouver, BC, Canada, 2011. ACM, pages 33–36. DOI: 10.1145/1979742.1979570.

[51] North, C., Dwyer, T., Lee, B., Fisher, D., Isenberg, P., Robertson, G., and Inkpen, K. Understanding multi-touch manipulation for surface computing. *Proc. of the 12th IFIP TC 13 International Conference on Human-Computer Interaction: Part II*, Uppsala, Sweden, 2009. Springer-Verlag, pages 236–249. DOI: 10.1007/978-3-642-03658-3_31. 74

[52] Olsen, K. A., Korfhage, R. R., Sochats, K. M., Spring, M. B., and Williams, J. G. Visualization of a document collection: the vibe system. *Inf. Process. Manage.*, 29(1), 1993, pages 69–81. DOI: 10.1016/0306-4573(93)90024-8. 21, 67

[53] Pak Chung, W., Hetzler, B., Posse, C., Whiting, M., Havre, S., Cramer, N., Anuj, S., Singhal, M., Turner, A., and Thomas, J. *IN-SPIRE InfoVis 2004 Contest Entry*, 2004. DOI: 10.1109/infvis.2004.37. 69

[54] Pearson, K. *On Lines and Planes of Closest Fit to Systems of Points in Space*, 1901. DOI: 10.1080/14786440109462720. 45

[55] Peck, S. M., North, C., and Bowman, D. A multiscale interaction technique for large, high-resolution displays. *Proc. of the 2009 IEEE Symposium on 3D User Interfaces*, 2009. IEEE Computer Society, pages 31–38. DOI: 10.1109/3dui.2009.4811202.

[56] Pike, W. A., Stasko, J., Chang, R., and O'Connell, T. A. The science of interaction. *Information Visualization*, 8(4), pages 263–274. DOI: 10.1057/ivs.2009.22. 12

[57] Pirolli, P. and Card, S. Sensemaking processes of intelligence analysts and possible leverage points as identified though cognitive task analysis. *Proc. of the 2005 International Conference on Intelligence Analysis, McLean, Virginia*, 2005, page 6. 2, 7, 8

[58] Plaisant, C., Grinstein, G., Scholtz, J., Whiting, M., O'Connell, T., Laskowski, S., Chien, L., Tat, A., Wright, W., Gorg, C., Zhicheng, L., Parekh, N., Singhal, K., and Stasko, J.

Evaluating visual analytics at the 2007 VAST symposium contest. *Computer Graphics and Applications, IEEE*, 28(2), 2008, pages 12–21. DOI: 10.1109/mcg.2008.27. 75

[59] Robinson, A. C. *Design for Synthesis in Geovisualization*. Ph.D. thesis, Pennsylvannia State University, University Park, PA, 2008.

[60] Rose, S., Engel, D., Cramer, N., and Cowley, W. Automatic keyword extraction from individual documents. *Text Mining*, 2010. John Wiley & Sons, Ltd, pages 1–20. DOI: 10.1002/9780470689646.ch1.

[61] Rüger, S. *Putting the User in the Loop: Visual Resource Discovery*. Springer Berlin/Heidelberg, 2006. DOI: 10.1007/11670834_1. 22

[62] Schiffman, S., Reynolds, L., and Young, F. *Introduction to Multidimensional Scaling: Theory, Methods, and Applications*. Academic Press, 1981. 49

[63] Shipman, F. and Marshall, C. Formality considered harmful: Experiences, emerging themes, and directions on the use of formal representations in interactive systems. *Comput. Supported Coop. Work*, 8(4), 1999, pages 333–352. DOI: 10.1023/a:1008716330212. 31, 74

[64] Shrinivasan, Y. B. and Wijk, J. J. V. Supporting the analytical reasoning process in information visualization. *Proc. of the 26th Annual SIGCHI Conference on Human Factors in Computing Systems*, Florence, Italy, 2008. ACM. DOI: 10.1145/1357054.1357247. 25

[65] Shupp, L., Andrews, C., Dickey-Kurdziolek, M., Yost, B., and North, C. Shaping the display of the future: The effects of display size and curvature on user performance and insights. *Human-Computer Interaction*, 24(1), 2009, pages 230–272. DOI: 10.1080/07370020902739429. 61

[66] Skupin, A. A cartographic approach to visualizing conference abstracts. *IEEE Computer Graphics and Applications*, 22, 2002, pages 50–58. DOI: 10.1109/38.974518. 3, 17, 19

[67] Spiegelhalter, D. and Lauritzen, S. Sequential updating of conditional probabilities on directed graphical structures. *Networks*, 20, 1990, pages 275–605. DOI: 10.1002/net.3230200507. 46

[68] Svensen, J. F. M. *GTM: The Generative Topographical Mapping*. Aston University, Birmingham, 1998.

[69] Thomas, J. J. and Cook, K. A. Illuminating the path. *IEEE Computer Society*, 2005. DOI: 10.1016/b978-044451014-3/50047-8. 2

[70] Tipping, M. E. and Bishop, C. M. Probabilistic principal component analysis. *Journal of the Royal Statistical Society, SeriesB: Statistical Methodology*, 61, 1999, pages 611–622. DOI: 10.1111/1467-9868.00196. 45, 46

[71] Torokhti, A. and Friedland, S. *Towards Theory of Generic Principal Component Analysis.* DOI: 10.1016/j.jmva.2008.07.005. 45

[72] Torres, R. S., Silva, C. G., Medeiros, C. B., and Rocha, H. V. Visual structures for image browsing. *Proc. of the 12th International Conference on Information and knowledge Management*, New Orleans, LA, 2003. ACM, pages 49–55. DOI: 10.1145/956863.956874. 20

[73] West, M. and Harrison, J. *Bayesian Forecasting and Dynamic Models (Springer Series in Statistics).* Springer, 1997. DOI: 10.1007/978-1-4757-9365-9. 46

[74] Wise, J. A., Thomas, J. J., Pennock, K., Lantrip, D., Pottier, M., Schur, A., and Crow, V. Visualizing the non-visual: Spatial analysis and interaction with information for text documents. *Readings in Information Visualization: Using Vision to Think*, 1999. Morgan Kaufmann Publishers Inc., pages 442–450. DOI: 10.1109/infvis.1995.528686. 19

[75] Wright, W., Schroh, D., Proulx, P., Skaburskis, A., and Cort, B. The sandbox for analysis: Concepts and methods. *CHI '06*, New York, NY, 2006. ACM, pages 801–810. DOI: 10.1145/1124772.1124890. 18

[76] Xing, E. P., Ng, A. Y., Jordan, M. I., and Russell, S. Distance metric learning, with application to clustering with side-information. *Advances in Neural Information Processing Systems 15*, 2002. MIT Press.

[77] Xu, R. and Wunsch, D. Survey of clustering algorithms. *IEEE Transactions on Neural Networks*, 16(3), 2005, pages 645–678. DOI: 10.1109/tnn.2005.845141. 19

[78] Yi, J. S., Kang, Y. A., Stasko, J., and Jacko, J. Toward a deeper understanding of the role of interaction in information visualization. *IEEE Transactions on Visualization and Computer Graphics*, 13(6), 2007, pages 1224–1231. DOI: 10.1109/tvcg.2007.70515. 11

[79] Yi, J. S., Melton, R., Stasko, J., and Jacko, J. A. Dust and magnet: Multivariate information visualization using a magnet metaphor. *Information Visualization*, 4(4), 2005, pages 239–256. DOI: 10.1057/palgrave.ivs.9500099. 21, 55, 67

[80] Yost, B., Haciahmetoglu, Y., and North, C. Beyond visual acuity: The perceptual scalability of information visualizations for large displays. *CHI 2007*, San Jose, California, 2007. ACM, pages 101–110. DOI: 10.1145/1240624.1240639.

[81] Russell, D. M., Stefik, M. J., Pirolli, P., and Card, S. K. The cost structure of sensemaking. In *Proc. of the INTERACT'93 and CHI'93 Conference on Human Factors in Computing Systems*, ACM, 1993, pages 269–276. DOI: 10.1145/169059.169209. 7, 8

[82] Klein, G., Moon, B., and Hoffman, R. Making sense of sensemaking 1: Alternative perspectives. *IEEE Intelligent Systems*, 21(4), 2006, pages 70–73. DOI: 10.1109/mis.2006.75. 7

[83] Klein, G., Moon, B., and Hoffman, R. Making sense of sensemaking 2: A macrocognitive model. *IEEE Intelligent Systems*, 21(5), 2006, pages 88–92. DOI: 10.1109/mis.2006.100. 7, 8, 9

[84] Tominski, C. Interaction for visualization. *Synthesis Lectures on Visualization*, 3(1), 2015, pages 1–107. DOI: 10.2200/s00651ed1v01y201506vis003. 12

[85] Van Wijk, J. J. The value of visualization. In *16th IEEE Visualization 2005 (VIS 2005)*, page 11. IEEE Computer Society, 2005. DOI: 10.1109/vis.2005.102. 9, 10

[86] Horvitz, E. Principles of mixed-initiative user interfaces. In *Proc. of the SIGCHI Conference on Human Factors in Computing Systems, CHI '99*, New York, NY, 1999. ACM, pages 159–166. DOI: 10.1145/302979.303030. 14

[87] Endert, A., Hossain, M. S., Ramakrishnan, N., North, C., Fiaux, P., and Andrews, C. The human is the loop: New directions for visual analytics. *Journal of Intelligent Information Systems*, Jan. 2014, pages 1–25. DOI: 10.1007/s10844-014-0304-9. 13, 14, 15

[88] Kirsh, D. The intelligent use of space. *Artif. Intell.*, 73, 1995, pages 31–68, DOI: 10.1016/0004-3702(94)00017-u. 17

[89] Dourish, P. *Where the Action is: The Foundations of Embodied Interaction*. MIT Press, Cambridge, 2001. 17

[90] Ruotsalo, T., Jacucci, G., Myllymäki, P., and Kaski, S. Interactive intent modeling: Information discovery beyond search. *Commun. ACM*, 58(1), Dec. 2014, pages 86–92. DOI: 10.1145/2656334. 20

[91] Ragan, E., Endert, A., Sanyal, J., and Chen, J. Characterizing provenance in visualization and data analysis: An organizational framework of provenance types and purposes. *Visualization and Computer Graphics, IEEE Transactions on*, 22(1), Jan 2016, pages 31–40. DOI: 10.1109/tvcg.2015.2467551. 25

[92] Kim, H., Choo, J., Park, H., and Endert, A. Interaxis: Steering scatterplot axes via observation-level interaction. *Visualization and Computer Graphics, IEEE Transactions on*, 22(1), Jan 2016, pages 131–140. DOI: 10.1109/tvcg.2015.2467615. 29, 56

[93] Attfield, S. J., Hara, S. K., and Wong, B. W. Sensemaking in visual analytics: Processes and challenges. *The Eurographics Association*, 2010, pages 1–6. 8

[94] Weick, K. E. *Sensemaking in Organizations*, volume 3. Sage, 1995. 8

[95] Sacha, D., Stoffel, A., Stoffel, F., Kwon, B. C., Ellis, G., and Keim, D. A. Knowledge generation model for visual analytics. *Visualization and Computer Graphics, IEEE Transactions on*, 20(12), 2014, pages 1604–1613. DOI: 10.1109/tvcg.2014.2346481. 10, 11

[96] Brown, E. T., Ottley, A., Zhao, H., Lin, Q., Souvenir, R., Endert, A., and Chang, R. *Finding Waldo: Learning about Users from their Interactions*, 2014. DOI: 10.1109/tvcg.2014.2346575. 13

[97] Brown, E. T., Chang, R., and Endert, A. Human-machine-learner interaction: The best of both worlds. *Human Centred Machine Learning Workshop at ACM CHI*, 2016. 14

[98] Dhar, V. Data science and prediction. *Communications of the ACM*, 56(12), 2013, pages 64–73. DOI: 10.1145/2500499. 3

[99] Sacha, D., Senaratne, H., Kwon, B. C., Ellis, G., and Keim, D. A. The role of uncertainty, awareness, and trust in visual analytics. *IEEE Transactions on Visualization and Computer Graphics*, 22(1), 2016, pages 240–249. DOI: 10.1109/tvcg.2015.2467591. 2

[100] Amershi, S., Fogarty, J., Kapoor, A., and Tan, D. S. Effective end-user interaction with machine learning. In *AAAI*, 2011. 12

[101] Amershi, S., Fogarty, J., and Weld, D. Regroup: Interactive machine learning for on-demand group creation in social networks. In *Proc. of the SIGCHI Conference on Human Factors in Computing Systems*, ACM, 2012, pages 21–30. DOI: 10.1145/2207676.2207680.

[102] Sacha, D., Sedlmair, M., Zhang, L., Lee, J. A., Weiskopf, D., North, S., and Keim, D. *Human-centered Machine Learning Through Interactive Visualization*, 2016. 12

[103] Lu, Y., Krüger, R., Thom, D., Wang, F., Koch, S., Ertl, T., and Maciejewski, R. Integrating predictive analytics and social media. In *Visual Analytics Science and Technology (VAST), 2014 IEEE Conference on*, IEEE, 2014, pages 193–202. DOI: 10.1109/vast.2014.7042495. 13

[104] Krause, J., Perer, A., and Ng, K. Interacting with predictions: Visual inspection of black-box machine learning models. In *Proc. of the 2016 CHI Conference on Human Factors in Computing Systems*, ACM, 2016, pages 5686–5697. DOI: 10.1145/2858036.2858529. 13

[105] Endert, A., Fiaux, P., and North, C. Unifying the Sensemaking Process with Semantic Interaction. *NSF Science of Interaction for Data and Visual Analytics Workshop*, 2012. 18, 23

[106] i2 Analyst's Notebook.

[107] Javed, W. and Elmqvist, N. ExPlates: Spatializing interactive analysis to scaffold visual exploration. In *Proc. of the 15th Eurographics Conference on Visualization, EuroVis '13*, pages 441–450, Aire- la-Ville, Switzerland, Switzerland, 2013. Eurographics Association. DOI: 10.1111/cgf.12131. 21

Author's Biography

ALEX ENDERT

Alex Endert is an Assistant Professor in the School of Interactive Computing at Georgia Tech. He directs the Visual Analytics Lab, where he and his students explore novel user interaction techniques for visual analytics. His lab often apply this fundamental research to applied domains including text analysis, intelligence analysis, cyber security, decision-making, and others. He is an active contributor to venues for human-computer interaction and information visualization (ACM CHI, IEEE VIS, IEEE TVCG, IEEE CG&A). He received his Ph.D. in Computer Science at Virginia Tech in 2012, advised by Dr. Chris North. In 2013, his work on Semantic Interaction was awarded the IEEE VGTC VPG Pioneers Group Doctoral Dissertation Award and the Virginia Tech Computer Science Best Dissertation Award.

Printed in the United States
by Baker & Taylor Publisher Services